POLYMER CHARACTERIZATION

POLYMER

CHARACTERIZATION

Laboratory Techniques and Analysis

by

Nicholas P. Cheremisinoff, Ph.D.

np   **NOYES PUBLICATIONS**
Westwood, New Jersey, U.S.A.

Library of Congress Catalog Card Number: 96010912
ISBN: 0-8155-1403-4
Printed in the United States

Published in the United States of America by
Noyes Publications
369 Fairview Avenue
Westwood, New Jersey 07675

10 9 8 7 6 5 4 3 2 1

Library of Congress Cataloging-in-Publication Data

Cheremisinoff, Nicholas P.
        Polymer characterization : laboratory techniques and analysis / by
    Nicholas P. Cheremisinoff.
            p.   cm.
        Includes index.
        ISBN 0-8155-1403-4
        1. Polymers--Analysis. I. Title.
    QD139.P6C483   1996
    620.1'92'0287--dc20   96-10912
                                                                  CIP

ABOUT THE AUTHOR

Nicholas P. Cheremisinoff is a private consultant to industry, academia, and government. He has nearly twenty years of industry and applied research experience in elastomers, synthetic fuels, petrochemicals manufacturing, and environmental control. A chemical engineer by trade, he has authored over 100 engineering textbooks and has contributed extensively to the industrial press. He is currently working for the United States Agency for International Development in Eastern Ukraine, where he is managing the Industrial Waste Management Project. Dr. Cheremisinoff received his B.S., M.S., and Ph.D. degrees from Clarkson College of Technology.

PREFACE

This volume provides an overview of polymer characterization test methods. The methods and instrumentation described represent modern analytical techniques useful to researchers, product development specialists, and quality control experts in polymer synthesis and manufacturing. Engineers, polymer scientists and technicians will find this volume useful in selecting approaches and techniques applicable to characterizing molecular, compositional, rheological, and thermodynamic properties of elastomers and plastics.

It is essential that both R&D laboratories as well as quality control functions be versed in the various techniques described in this book in order to properly design their products and to ensure the highest quality polymers on the market place. This volume is particularly useful in introducing standard polymer characterization laboratory techniques to technicians and engineers beginning their careers in polymer manufacturing and product development areas.

A large portion of this volume is comprised of appendices providing definitions of testing and product characterization terms. These sections are intended to provide the reader with a practical source of fundamental information. The author wishes to extend gratitude to Linda Jastrzebski for her assistance in organizing and typesetting this book.

Nicholas P. Cheremisinoff

NOTICE

To the best of our knowledge the information in this publication is accurate; however, the Publisher does not assume any responsibility or liability for the accuracy or completeness of, or consequences arising from, such information. This book is intended for informational purposes only. Mention of trade names or commercial products does not constitute endorsement or recommendation for use by the Publisher. Final determination of the suitability of any information or product for use contemplated by any user, and the manner of that use, is the sole responsibility of the user. We recommend that anyone intending to rely on any recommendation of materials or procedures mentioned in this publication should satisfy himself as to such suitability, and that he can meet all applicable safety and health standards.

CONTENTS

1 CHROMATOGRAPHIC TECHNIQUES

CHROMATOGRAPHY FOR ANALYTICAL ANALYSES

Chromatography may be defined as the separation of molecular mixtures by distribution between two or more phases, one phase being essentially two-dimensional (a surface) and the remaining phase, or being a bulk phase brought into contact in a counter-current fashion with the two-dimensional phase. Various types of physical states of chromatography are possible, depending on the phases involved.

Chromatography is divided into two main branches. One branch is gas chromatography, the other is liquid chromatography. Liquid chromatography can be further subdivided as shown in Figure 1.

The sequence of chromatographic separation is as follows: A sample is placed at the top of a column where its components are sorbed and desorbed by a carrier. This partitioning process occurs repeatedly as the sample moves towards the outlet of the column. Each solute travels at its own rate through the column, consequently, a band representing each solute will form on the column. A detector attached to the column's outlet responds to each band. The output of detector response versus time is called a chromatogram. The time of emergence identifies the component, and the peak area defines its concentration, based on calibration with known compounds.

GAS CHROMATOGRAPHY

General

If the moving phase is a gas, then the technique is called gas chromatography (GC). In gas chromatography the sample is usually

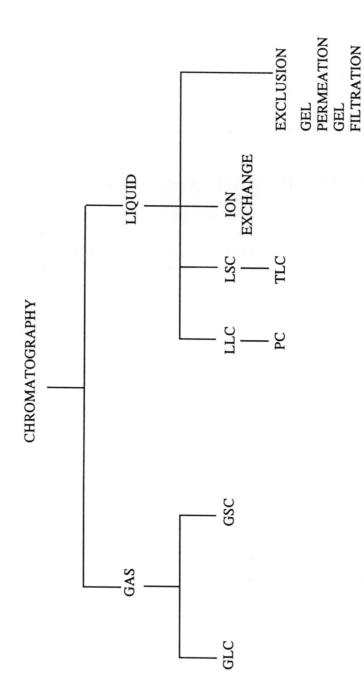

Figure 1. Shows types of chromatographic operations.

injected at high temperature to ensure vaporization. Obviously, only materials volatile at this temperature can be analyzed.

Types of GC

If the stationary phase is a solid, the technique is referred to as gas-solid chromatography. The separation mechanism is principally one of adsorption. Those components more strongly adsorbed are held up longer than those which are not.

If the stationary phase is a liquid, the technique is referred to as gas-liquid chromatography and the separation mechanisms is principally one of partition (solubilization of the liquid phase).

Gas chromatography has developed into one of the most powerful analytical tools available to the organic chemist. The technique allows separation of extremely small quantities of material (10^{-6} grams).

The characterization and quantitation of complex mixtures can be accomplished with this process. The introduction of long columns, both megabore and capillary, produces a greater number of theoretical plates increasing the efficiency of separation beyond that of any other available technique. The technique is applicable over a wide range of temperatures (-40-350°C) making it possible to chromatograph materials covering a wide range of volatiles. The laboratory uses packed columns along with megabore and capillary. In this way the broadest range of chromatographic problems can be addressed.

The detector used to sense and quantify the effluent provides the specificity and sensitivity for the analytical procedure. Table 1 summarizes significant detector characteristics.

LIQUID CHROMATOGRAPHY

General

If the moving phase is a liquid, then the technique is called liquid chromatography (LC). In liquid chromatography the sample is first dissolved in the moving phase and injected at ambient temperature. Thus there is no volatility requirement for samples. However, the sample must dissolve in the moving phase. Note that LC has an important advantage over GC: The solubility requirement can usually be met by

TABLE 1.
SUMMARY OF DETECTOR CHARACTERISTICS

Detector	Principle of Operation	Selectivity	Sensitivity	Linear Range	MDQ[1]	Stability
Thermal Conductivity	Measures thermal conductivity of gas	Universal	6×10^{-10}	10^4	10^{-5} gm of CH_4 per vol. of detector effluent	Good
Flame Ionization	$H_2 - O_2$ Flame	Responds to organic compounds, not to H_2O or fixed gases	9×10^{-3} for alkane	10^7	2×10^{-11} gm for alkane	Excellent
Electron Capture	$N2 + B \rightarrow e^-$ $e^- + Sample \rightarrow$	Responds to electron adsorbing compounds, e.g., halogen	2×10^{-14} for CCL_4	10^5		Good
Hall Electrolytic Conductivity Detector		In halogen mode responds to halogens		10^0	1×10^{13} g cl/sec	Poor

[1]Minimum detectable quantity.

changing the moving phase. The volatility requirement is not so easily overcome.

Types of LC

There are four kinds of liquid chromatography, depending on the nature of the stationary phase and the separation mechanism:

- *Liquid/Liquid Chromatography (LLC)*--is partition chromatography or solution chromatography. The sample is retained by partitioning between mobile liquid and stationary liquid. The mobile liquid cannot be a solvent for the stationary liquid. As a subgroup of liquid/liquid chromatography there is paper chromatography.
- *Liquid/Solid Chromatography (LSC)*--is adsorption chromatography. Adsorbents such as alumina and silica gel are packed in a column and the sample components are displaced by a mobile phase. Thin layer chromatography and most open column chromatography are considered liquid/solid chromatography.
- *Ion-Exchange Chromatography*--employs zeolites and synthetic organic and inorganic resins to perform chromatographic separation by an exchange of ions between the sample and the resins. Compounds which have ions with different affinities for the resin can be separated.
- *Exclusion Chromatography*--is another form of liquid chromatography. In the process a uniform nonionic gel is used to separate materials according to their molecular size. The small molecules get into the polymer network and are retarded, whereas larger molecules cannot enter the polymer network and will be swept our of the column. The elution order is the largest molecules first, medium next and the smallest sized molecules last. The term "gel permeation chromatography" has been coined for separations polymers which swell in organic solvent.

The trend in liquid chromatography has tended to move away from open column toward what is called high pressure liquid chromatography (HPLC) for analytical as well as preparative work. The change in technique is due to the development of high sensitivity, low dead volume detectors. The result is high resolution, high speed, and better sensitivity liquid chromatography.

Type of Information Obtained

Form

The output of a chromatographic instrument can be of two types:

- A plot of areas retention time versus detector response. The peak areas represent the amount of each component present in the mixture.
- A computer printout giving names of components and the concentration of each in the sample.

Units

The units of concentration are reported in several ways:

- Weight percent or ppm by weight.

Volume percent or ppm by volume.

- Mole percent.

Sample

Size--A few milligrams is usually enough for either GC or LC.

State

1. For GC, the sample can be gas, liquid, or solid. Solid samples are usually dissolved in a suitable solvent; both liquid or solid samples must volatilize at the operating temperature.

2. For LC, samples can be liquid or solid. Either must be soluble in moving phase.

Advantages

Gas Chromatography

- Moderately fast quantitative analyses (0.5-1.5 hours per sample).
- Excellent resolution of various organic compounds.
- Not limited by sample solubility.
- Good sensitivity.
- Specificity.

Liquid Chromatography

- Separation of high boiling compounds.
- Not limited by sample volatility.
- Moving phase allows additional control over separation.

Disadvantages

Gas Chromatography

- Limited by sample volatility.

Liquid Chromatography

- Less sensitive than GC.
- Detectors may respond to solvent carrier, as well as to sample.

Interferences

Interferences in chromatography can generally be overcome by finding the right conditions to give separation. However, this might be costly, since development of separations is largely a trial-and-error process.

GPC/DRI

For common linear homopolymers, such as PIB, PE, PS..., GPC analysis can be performed with a single DRI detector. Figure 2 shows the basic component of a GPC/DRI System. Most often, a PS calibration curve is generated from narrow molecular weight PS standards (Figure 3), which can then be converted to the desired polymer (i.e., PIB, EP...) if the appropriate calibration constants are available. These constants, known as the Mark-Houwink parameters or k and alpha, are used to calculate the intrinsic viscosity of the polymer as a function of molecular weight (which is needed to relate the size of one type of polymer to another). If the Mark-Houwink parameters are not available, the molecular weights can be used for relative comparison but will not be correct on an absolute basis. If the sample is branched, the molecular weights will be biased low, and a secondary detector (LALLS or VIS) is needed for accurate results.

GEL PERMEATION CHROMATOGRAPHY

Gel Permeation Chromatography (GPC), also known as Size Exclusion Chromatography (SEC), is a technique used to determine the average molecular weight distribution of a polymer sample. Using the appropriate detectors and analysis procedure it is also possible to obtain qualitative information on long chain branching or determine the composition distribution of copolymers.

As the name implies, GPC or SEC separates the polymer according to size or hydrodynamic radius. This is accomplished by injecting a small amount of (100-400 μl) of polymer solution (0.01-0.6%) into a set of columns that are packed with porous beads. Smaller molecules can penetrate the pores and are therefore retained to a greater extent than the larger molecules which continue down the columns and elute faster. This process is illustrated in Figure 4.

One or more detectors is attached to the output of the columns. For routine analysis of linear homopolymers, this is most often a Differential Refractive Index (DRI) or a UV detector. For branched or copolymers, however, it is necessary to have at least two sequential detectors to determine molecular weight accurately. Branched polymers can be analyzed using a DRI detector coupled with a "molecular weight sensitive" detector such as an on-line viscometer (VIS) or a low-angle laser light scattering (LALLS) detector. The compositional distribution of copolymers, i.e., average composition as a function of molecular size, can be determined using a DRI detector coupled with a selective detector

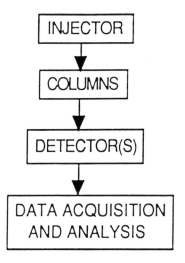

Figure 2. Basic components of GPC/DRI.

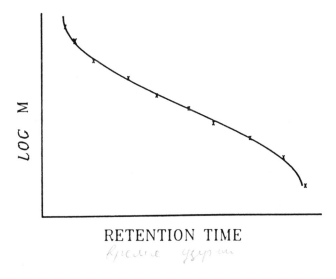

Figure 3. Typical calibration curve using a polystyrene (PS) standard.

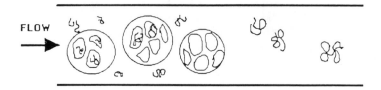

Figure 4. Polymer solution flow through GPC column.

such as UV or FTIR. It is important to consider the type of polymer and information that is desired before submitting a sample. The following outline describes each instrument that is currently available.

GPC/DRI/LALLS

One can use two instruments with sequential LALLS and DRI detectors. The unit is operated using TCB at 135°C and is used to analyze PE, EP, and PP samples. The other, operating at 60°C, is for butyl type polymers which dissolve in TCB at lower temperatures. The data consists of two chromatograms, plots of detector mV signal (LALLS and DRI) versus retention time. The DRI trace corresponds to the concentration profile whereas the LALLS signal is proportional to concentration *M, resulting in more sensitivity at the high molecular weight end. An example of the output is shown in Figure 5 for polyethylene NBS 1476. The LALLS trace shows a peak at the high molecular weight end (low retention time) which is barely noticeable on the DRI trace. This suggests a very small amount of high molecular weight, highly branched material. This type of bimodal peak in the DRI trace is often seen in branched EP (ethylene-propylene polymers) or LDPE (low density polyethylene) samples. The report consists of two result pages, one from the DRI calibration curve as described above, and the second from the LALLS data. An example of a report page is shown in Figure 6. At the top of the page should be a file name and date of analysis. The header also includes a description of the method and detector type, which in this case is the DRI detector and EP calibration curve. Following the header are the parameters integration (i.e., start and end times for integration and baseline) and a slice report (i.e., cumulative weight percent and molecular weight as a function of retention time). This section gives details about the distribution, such as the range of molecular weights for the sample and the fraction of polymer above a particular molecular weight. At the bottom of the page is a summary of the average molecular weights, whereas Z denotes the Z average molecular weight or Mz, etc.

For a linear polymer (if all the calibration constants are known), the molecular weights from both pages should agree within 10%. A LALLS report that gives higher molecular weights than the DRI suggests that the sample is branched, and the values from the LALLS report should be used (again, assuming that the calibration constants are correct). Occasionally, some of the sample, gel or insolubles, is filtered out during the sample preparation and analysis. The percentage should be indicated on the report.

GPC/DRI/VIS

GPC with an on-line viscometer can be used instead of a LALLS detector to analyze branched polymers. In this case the intrinsic viscosity is measured so that the Mark-Houwink parameters are not needed. It is complementary to the LALLS instrument in intrinsic viscosities.

GPC/DRI/UV

The UV detector is used to analyze chromophores. Its most common use is for graft or block copolymers containing PS or PMS. The data from this instrument consists of two chromatograms, the UV and DRI traces. An example is shown in Figure 7 for an EP-g-PS coplymer (peak 1) with risidual PS homopolymer (peak 2). The UV absorption relative to the DRI signal corresponds to the copolymer composition, which is why the relative UV absorption is higher for the pure PS in peak 2. The results report consists of two pages. One is the molecular weight report from the DRI calibration curve as described above. Note that the molecular weights are reported as if the sample is a homopolymer not copolymer. The other page using the UV data gives an effective extinction coefficient E' which is the UV/DRI ratio. A higher E'

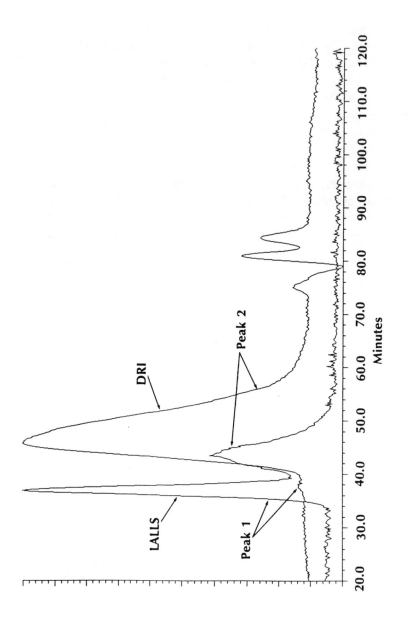

Figure 5. Example of output from a LALLS detector.

02029:1	NBS-1476-8		30 Mar 1990	1:24:03	Page 1

Analyzed   3 Apr 1990   10:57:15

High Speed GPC - Calibration Curve 206   INST B EP TCB
Detector - DRI

Peak Parameters - Time, min		Mvolts	Baseline Height	Time
Start	36.00	05.0	4.96	35.00
Max	46.00	63.5		
Finish	66.00	06.3	6.26	67.00

Peak Area, mv-sec = 33373.   Molecular Weight at Peak Max   79442

Time, min	Height, mv	dWM/dLM	Cum Wt Pct	Mol Pct	Cum Mol Pct	Mol Wt
36.00	00.000	0.00	00.00	0.00	00.00	5050122
37.20	01.149	0.01	00.15	0.00	00.00	2699310
38.40	00.800	0.01	00.36	0.00	00.00	1501013
39.60	00.934	0.01	00.52	0.00	00.01	0866453
40.80	04.760	0.03	01.09	0.02	00.03	0518072
42.00	16.159	0.11	03.44	0.15	00.18	0320161
43.20	31.938	0.27	08.82	0.53	00.70	0204054
44.40	49.795	0.50	18.04	1.38	02.08	0133833
45.60	57.835	0.70	30.00	2.65	04.73	0090134
46.80	56.523	0.77	42.42	4.01	08.74	0062195
48.00	53.013	0.78	54.20	5.43	14.16	0043878
49.20	46.600	0.75	64.87	6.87	21.04	0031579
50.40	38.402	0.67	73.90	7.98	29.01	0023134
51.60	30.066	0.56	81.14	8.63	37.65	0017214
52.80	22.703	0.45	86.67	8.80	46.44	0012982
54.00	17.144	0.35	90.85	8.75	55.19	0009900
55.20	12.479	0.27	93.97	8.52	63.71	0007619
56.40	08.370	0.19	96.12	7.59	71.30	0005904
57.60	05.380	0.13	97.54	6.43	77.73	0004597
58.80	03.529	0.09	98.47	5.39	83.11	0003588
60.00	02.208	0.05	99.05	4.38	87.49	0002801
61.20	01.595	0.04	99.44	3.69	91.17	0002183
62.40	01.085	0.02	99.70	3.25	94.43	0001694
63.60	00.517	0.01	99.87	2.54	96.97	0001307
64.80	00.360	0.01	99.95	1.84	98.82	0001000
66.00	00.053	0.00	100.00	1.18	100.00	0000757

Average Mol Wts   Ratios of Averages   Time Int Std Peak, min

(Z-1)	=	2080106	(Z-1)/Z	=	04.417	Expected	85.20
Z	=	0470912	(Z-1)/WT	=	23.598	Actual	84.20
WT	=	0088146	Z/WT	=	05.342		
VIS	=	0073154	WT/VIS	=	01.205		
AVIS	=	0.726	VIS/NO	=	03.144		
NO	=	0023266	WT/NO	=	03.789	Intrinsic Viscosity	0.993

Figure 6. Example of a report page.

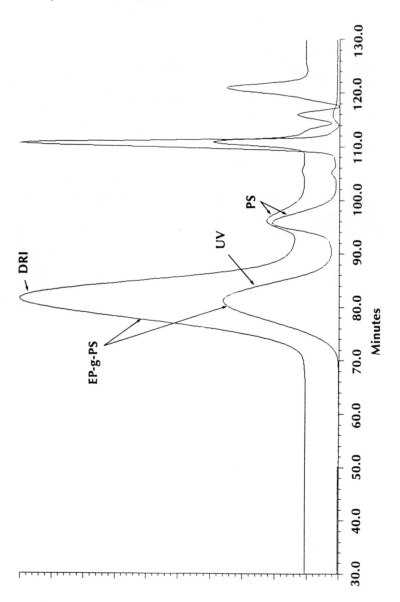

Figure 7. Example of UV detector output.

indicates a higher composition of the UV active chromophore (for example, more PS in the graft copolymer). This technique is also used to determine the compositional distribution of ENB (ethylidene norbonene) in EPDM, i.e., whether the ENB is evenly distributed across the molecular weight distribution or concentrated in the low or high molecular weight end. The GPC/DRI/UV instrument can be used to analyze samples that dissolve in THF at 30-45°C.

GPC/DRI/FTIR

The GPC/DRI/FTIR instrument is complementary to the UV detector for compositional distribution. It runs at 135°C in TCB and can be used for EP analysis. Typical applications include ethylene content as a function of molecular weight, maleic anhydride content in maleated EP, or PCL content in caprolactone-g-EP copolymers. The FTIR detector is off-line so that 5-10 fractions of the eluant are collected on KBr plates and analyzed. This procedure gives calibration of IR absorption bands. This method is much more labor intensive than the other techniques and should be used with discretion.

Submitting Samples

Samples should be weighed out (typically 30-120 μg) in bottles. The submitter should check which is the appropriate amount for a particular test. The sample should be labeled with the contents, exact amount of polymer, and test type. Any other information, such as expected molecular weight range, ENB or other monomer content, dissolution temperature..., is helpful for optimizing the analysis. Typically, a single GPC run takes approximately 2½ hours, except for GPC/FTIR which can take five hours for the fractionation and additional time for the FTIR data acquisition.

2 THERMAL ANALYSIS

GENERAL PRINCIPLES OF OPERATION

Thermal analysis refers to a variety of techniques in which a property of a sample is continuously measured as the sample is programmed through a predetermined temperature profile. Among the most common techniques are thermal gravimetric analysis (TA) and differential scanning calorimetry (DSC).

In TA the mass loss versus increasing temperature of the sample is recorded. The basic instrumental requirements are simple: a precision balance, a programmable furnace, and a recorder (Figure 1). Modern instruments, however, tend to be automated and include software for data reduction. In addition, provisions are made for surrounding the sample with an air, nitrogen, or an oxygen atmosphere.

In a DSC experiment the difference in energy input to a sample and a reference material is measured while the sample and reference are subjected to a controlled temperature program. DSC requires two cells equipped with thermocouples in addition to a programmable furnace, recorder, and gas controller. Automation is even more extensive than in TA due to the more complicated nature of the instrumentation and calculations.

A thermal analysis curve is interpreted by relating the measured property versus temperature data to chemical and physical events occurring in the sample. It is frequently a qualitative or comparative technique.

In TA the mass loss can be due to such events as the volatilization of liquids and the decomposition and evolution of gases from solids. The onset of volatilization is proportional to the boiling point of the liquid. The residue remaining at high temperature represents the percent ash content of the sample. Figure 2 shows the TA spectrum of calcium oxalate as an example.

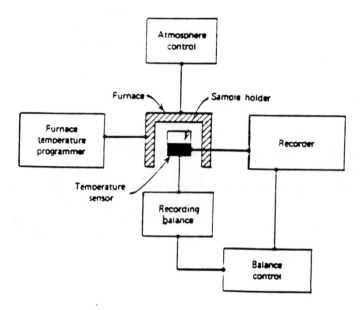

Figure 1. Typical components of a TA instrument.

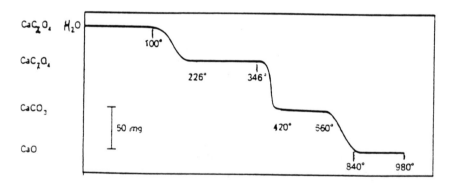

Figure 2. Shows the TA spectrum of calcium oxalate.

In DSC the measured energy differential corresponds to the heat content (enthalpy) or the specific heat of the sample. DSC is often used in conjunction with TA to determine if a reaction is endothermic, such as melting, vaporization and sublimation, or exothermic, such as oxidative degradation. It is also used to determine the glass transition temperature of polymers. Liquids and solids can be analyzed by both methods of thermal analysis. The sample size is usually limited to 10-20 mg.

Thermal analysis can be used to characterize the physical and chemical properties of a system under conditions that simulate real world applications. It is not simply a sample composition technique.

Much of the data interpretation is empirical in nature and more than one thermal method may be required to fully understand the chemical and physical reactions occurring in a sample.

Condensation of volatile reaction products on the sample support system of a TA can give rise to anomalous weight changes.

THERMAL ANALYSIS OF POLYMERS

A simple example of the relationship between "structure" and "properties" is the effect of increasing molecular weight of a polymer on its physical (mechanical) state; a progression from an oily liquid, to a soft viscoelastic solid, to a hard, glassy elastic solid. Even seemingly minor rearrangements of atomic structure can have dramatic effects as, for example, the atactic and syndiotactic stereoisomers of polypropylene-- the first being a viscoelastic amorphous polymer at room temperature while the second is a strong, fairly rigid plastic with a melting point above 160°C. At high thermal energies conformational changes via bond rotations are frequent on the time scale of typical processing operations and the polymer behaves as a liquid (melt). At lower temperatures the chains solidifies by either of two mechanisms: by ordered molecular packing in a crystal lattice, *crystallization*, or by a gradual freezing out of long range molecular motions, *vitrification*. These transformations, which define the principal rheological regimes of mechanical behavior: the melt, the rubbery state, and the semicrystalline and glassy amorphous solids, are accompanied by transitions in thermodynamic properties at the glass transition temperature, the crystalline melting, and the crystallization temperatures.

Thermal analysis techniques are designed to measure the above mentioned transitions both by measurements of heat capacity and mechanical modulus (stiffness).

Differential Scanning Calorimetry (DSC)

The DSC measures the power (heat energy per unit time) differential between a small weighed sample of polymer (ca. 10 mg) in a sealed aluminum pan referenced to an empty pan in order to maintain a zero temperature differential between them during programmed heating and cooling temperature scans.   The technique is most often used for characterizing the T_g, T_m, T_c, and heat of fusion of polymers (Figure 3). The technique can also be used for studying the kinetics of chemical reactions, e.g., oxidation and decomposition.   The conversion of a measured heat of fusion can be converted to a % crystallinity provided, of course, the heat of fusion for the 100% crystalline polymer is known.

Thermogravimetric Analysis (TGA)

TGA makes a continuous weighing of a small sample (ca 10 mg) in a controlled atmosphere (e.g., air or nitrogen) as the temperature is increased at a programmed linear rate.   The thermogram shown in Figure 4 illustrates weight losses due to desorption of gases (e.g., moisture) or decomposition (e.g., HBr loss from halobutyl, CO_2 from calcium carbonate filler).   TA is a very simple technique for quantitatively analyzing for filler content of a polymer compound (e.g., carbon black decomposed in air but not nitrogen).   While oil can be readily detected in the thermogram it almost always overlaps with the temperature range of hydrocarbon polymer degradation.   The curves cannot be reliably deconvoluted since the actual decomposition range of a polymer in a polymer blend can be affected by the sample morphology.

Thermomechanical Analysis (TMA)

TMA consists of a quartz probe which rests on top of a flat sample (a few mms square) in a temperature controlled chamber.   When setup in neutral buoyancy (with 'flat probe') then as the temperature is increased the probe rises in direct response to the expansion of the sample yielding

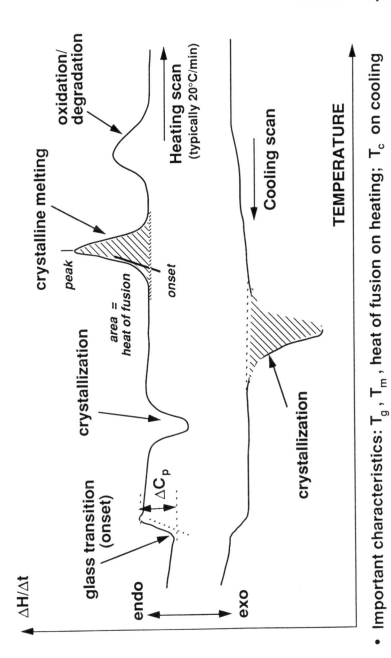

Figure 3. Illustrates typical polymer DSC thermograms.

• **Important characteristics: T_g, T_m, heat of fusion on heating; T_c on cooling**

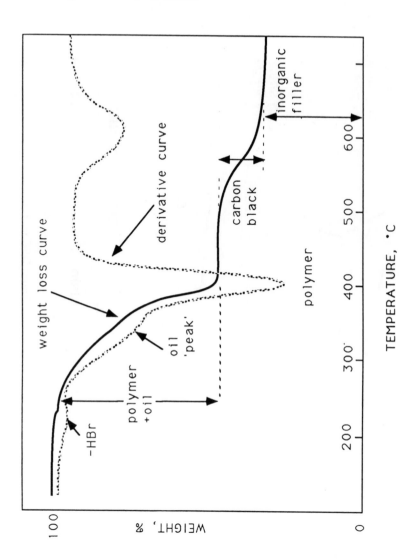

Figure 4. Shows a TA thermogram of an elastomer compound.

thermal expansion coefficient versus temperature scans. Alternatively, with the 'penetration probe' under dead loading a thermal softening profile is obtained (penetration distance versus temperature). Although this is a simple and versatile experiment, it gives only a semi-quantitative indication of mechanical modulus versus temperature. The DMTA, described below, gives an absolute modulus measurement.

Dynamic Mechanical Thermal Analysis (DMTA)

DMTA is a measurement of the dynamic moduli (in-phase and out-of-phase) in an oscillatory mechanical deformation experiment during a programmed temperature scan at controlled frequency. Thermograms are usually plotted to show elastic modulus, E, and tan ó versus temperature (Figure 5). The peak of the tan ó is a particularly discriminatory measure of T_g, although this is the center of the relaxation whereas in the DSC experiment the onset temperature of the T_g relaxation is usually reported. In such a case the DSC T_g will be lower than that for DMTA by an amount that varies with the specific polymer. There is, in addition, a frequency effect which puts the mechanical (ca. 1 Hz) T_g about 17°C higher than that for a DSC measurement (ca. 0.0001Hz) for an assumed activation energy of 400 kJ/mole (typical for polymer T_g). The DMTA has a frequency multiplexing capability which can be used for calculating activation energies using time-temperature superposition software.

The temperature range of the DMTA is from -150°C to 300°C and frequencies from 0.033 to 90 Hz. The sample size for the usual flexural test mode is 1 mm x 10 mm x 40 mm; slightly less sample is required in the parallel plate shear mode.

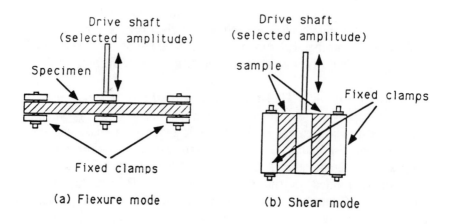

Dynamic Mechanical Thermal Analyzer

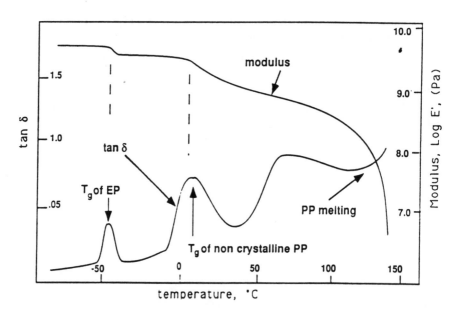

Figure 5 (a) and (b).   Show details of DMTA. The DMTA plot is of an EP/PP blend.

3 MICROSCOPY FOR
POLYMER CHARACTERIZATION

GENERAL INFORMATION

This chapter provides general information on the use of microscopic techniques for polymer characterization. For polymer blends a minimum domain size of 1 μm can be examined in the optical microscope using one or more of the following techniques. A schematic of a typical optical microscope is shown in Figure 1.

1. *Phase contrast*--Thin sections (100-200 nm) in thickness (and having refractive indices which differ by approximately .005) are supported on glass slides and examined "as is" or with oil to remove microtoming artifacts, e.g., determination of the number of layers in coextruded films, dispersion of fillers, and polymer domain size. (Figures 2 and 3)

2. *Polarized light*--Is used if one of the polymer phases is crystalline or for agglomeration of inorganic filters, (e.g., nylon/EP blends and fillers such as talc. (Figure 4)

3. *Incident*--Is used to examine surfaces of bulk samples, e.g., carbon black dispersion in rubber compounds. (Figure 5)

4. *Bright field*--Mainly used to examine thin sections of carbon black loaded samples, e.g., carbon black dispersion in thin films of rubber compounds.

When the domain size is in the range of < 1 μm to 10nm, scanning electron microscopy (SEM) and/or transmission electron microscopy

(TEM) are necessary. A schematic of a scanning electron microscope is shown in Figure 6.

Samples in the SEM can be examined "as is" for general morphology, as freeze fractured surfaces or as microtome blocks of solid bulk samples. Contrast is achieved by any one or combination of the following methods:

1. *Solvent etching*--When there exists a large solubility difference in a particular solvent of the polymers being studied, e.g., PP/EP blends.

2. *O_sO_4 Staining*--There exists at least 5% unsaturation in the polymers being investigated, e.g., NR/EPDM, BIIR/Neoprene. (Figure 7)

3. *RuO_4 Staining*--When there is no solubility differences or unsaturation this possibility is explored, e.g., knit explored line between two DVA's (dynamic vulcanized alloys). (Figure 8)

In addition, the SEM can be used to study liquids or temperature sensitive polymers on a Cryostage.

The SEM is also used to do X-ray/elemental analysis. This technique is qualitative. X-ray analysis and mapping of the particular elements present is useful for the identification of inorganic fillers and their dispersion in compounds as well as inorganic impurities in gels or on surfaces and curatives, e.g., aluminum, silicon, or sulfur in rubber compounds and Cl and Br in halobutyl blends. (Figure 9)

TEM (schematic shown in Figure 10) is used whenever a more in-depth study (when domain sizes are less than 1 micron or so) is required on polymer phase morphologies such as dynamically vulcanized alloys (Figure 11) and Nylon/EP (Figure 12) filler location as in carbon black in rubber compounds (Figure 13) and also in the morphology of block copolymers (Figure 14). Thin sections are required and take anywhere from one hour to one day per sample depending on the nature of the sample. They must be ~ 100 nm in thickness and are prepared usually by microtoming with a diamond knife at near liquid nitrogen temperatures (-150°C). The same contrasting media for SEM apply to TEM. In addition, PIB backbone polymers scission and evaporate in the TEM which helps locate these polymers domains in blends.

NON-ROUTINE TECHNIQUES

- Solvent casting when microtoming is not desirable as a method of sample preparation.

- *STEM*--Used for elemental composition study in thin films when better resolution is required than X-ray analysis in the SEM on bulk samples.

- *Cryostage - SEM*--To study liquid samples at low temperatures, e.g., butyl slurry.

- *Fluorescence microscope*--Useful in examining polymer/asphalt blends or any sample which is fluorescent.

- *OM/Hot stage*--To observe melting point of either an impurity or other moiety in a compound.

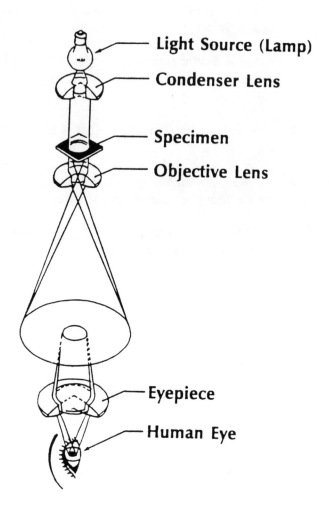

Light Source (Lamp)

Condenser Lens

Specimen

Objective Lens

Eyepiece

Human Eye

Optical Microscope

Figure 1. Schematic of an optical microscope.

Figure 2. Light microscopy phase contrast films.

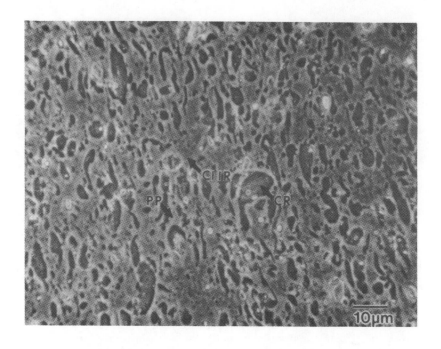

CIIR = Grey Areas

PP = White Areas

Neoprene (CR) = Dark Areas

Figure 3.   Light microscopy phase contrast polymer domains chlorobutyl / polypropylene / neoprene blend (CIIR/PP/CR).

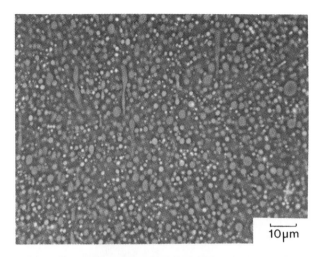

**EP is Light,
Dispersed Phase**

**Nylon is Dark
Matrix**

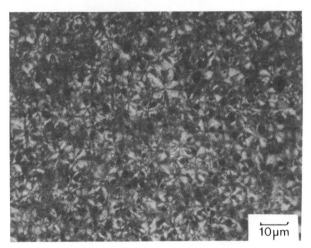

**Polarized Light
(Shows Spherultic Structures)**

Figure 4. Light microscopy phase contrast nylon/EP blends.

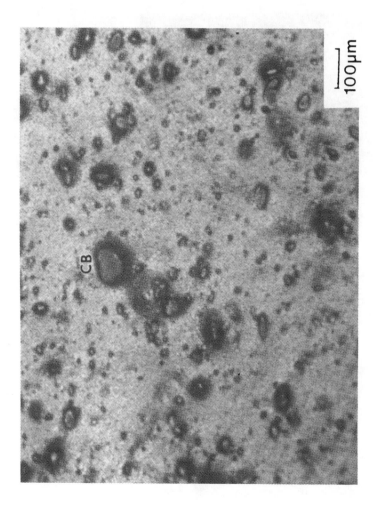

Figure 5. Light microscopy incident light carbon black dispersion in rubber.

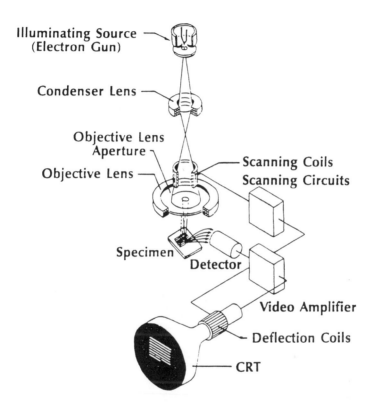

Figure 6. Schematic of a scanning electron microscope (SEM).

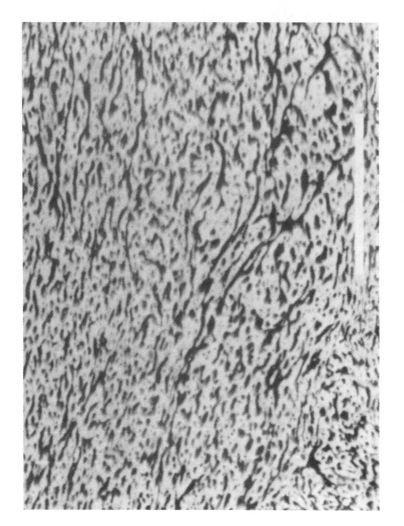

Neoprene is the Light Phase

Figure 7.   Enhancing SEM contrast in blends by osmium tetroxide staining bromobutyl/neoprene blend.

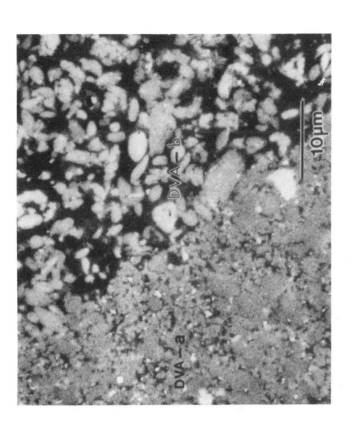

Figure 8.   SEM–ruthenium tetroxide stained knit line between two DVA's

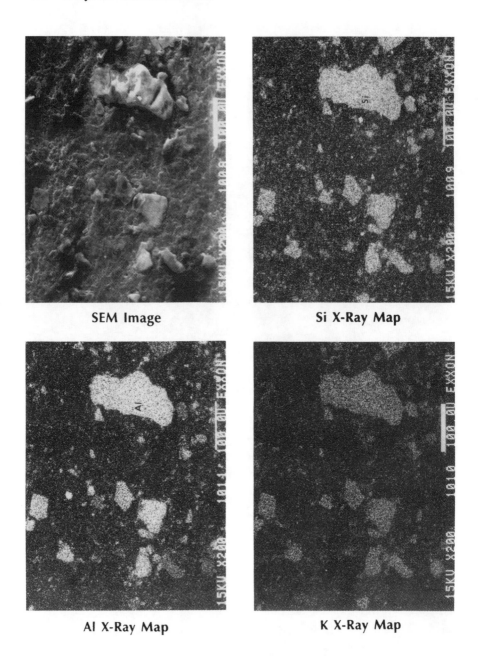

SEM Image

Si X-Ray Map

Al X-Ray Map

K X-Ray Map

Figure 9. X-ray mapping of surface impurity.

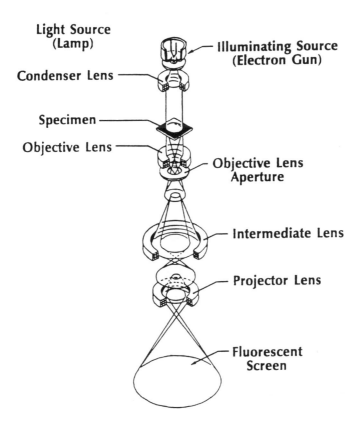

Figure 10. Schematic of a transmission electron microscope (TEM).

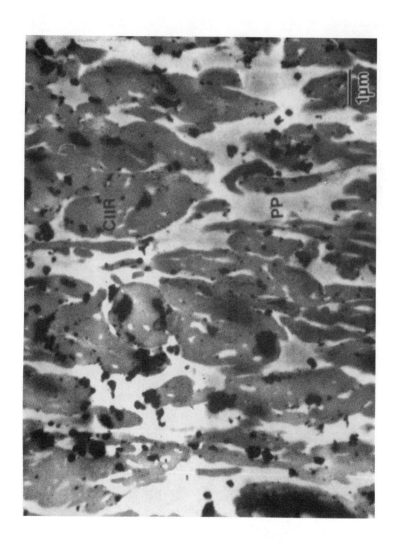

Figure 11.   TEM–ruthenium tetroxide stained polymer domains in a DVA compound.

Figure 12. Phase morphology in a nylon/EP-MA blend by TEM.

Figure 13.   Location of carbon black in a blend of chlorobutyl and natural rubber & EPDM.

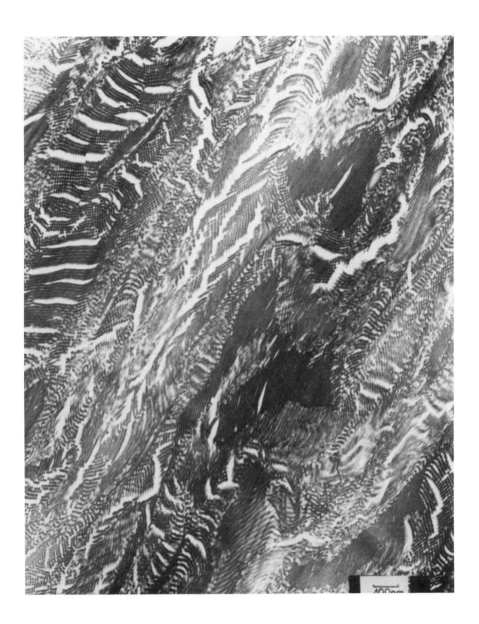

Figure 14. REM-Ruthenium tetroxide stained graft copolymer.

4 ELEMENTAL AND STRUCTURAL CHARACTERIZATION TESTS

ATOMIC ABSORPTION SPECTROSCOPY

In atomic absorption spectrometry (AA) the sample is vaporized and the element of interest atomized at high temperatures. The element concentration is determined based on the attenuation or absorption by the analyte atoms, of a characteristic wavelength emitted from a light source. The light source is typically a hollow cathode lamp containing the element to be measured. Separate lamps are needed for each element. The detector is usually a photomultiplier tube. A monochromator is used to separate the element line and the light source is modulated to reduce the amount of unwanted radiation reaching the detector.

Conventional AA instruments (Figure 1) use a flame atomization system for liquid sample vaporization. An air-acetylene flame (2300°C) is used for most elements. A higher temperature nitrous oxide-acetylene flame (2900°C) is used for more refractory oxide forming elements. Electrothermal atomization techniques such as a graphite furnace can be used for the direct analysis of solid samples.

Atomic absorption is used for the determination of ppm levels of metals. It is not normally used for the analysis of the light elements such as H, C, N, 0, P and S, halogens, and noble gases. Higher concentrations can be determined by prior dilution of the sample. AA is not recommended if a large number of elements are to be measured in a single sample.

Although AA is a very capable technique and is widely used worldwide, its use in recent years has declined in favor of ICP and XRF methods of analysis. The most common application of AA is for the determination of boron and magnesium in oils.

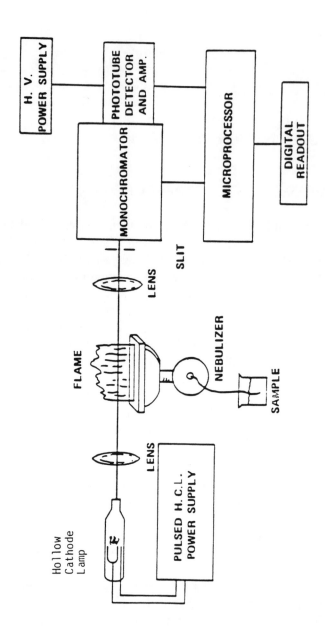

Figure 1. Illustrates major components of atomic absorption spectrometer.

Conventional AA instruments will analyze liquid samples only. Dilute acid and xylene solutions are common. The volume of solution needed is dependent on the number of elements to be determined.

AA offers excellent sensitivity for most elements with limited - interferences. For some elements sensitivity can be extended into the sub-ppb range using flameless methods. The AA instruments are easy to operate with "cookbook methods" available for most elements.

Conventional AA use a liquid sample. The determination of several elements per sample is slow and requires larger volumes of solution due to the sequential nature of the method.   Chemical and ionization interferences must be corrected by modification of the sample solution.

Chemical interferences arise from the formation of thermally stable compounds such as oxides in the flame.   The use of electrothermal atomization, a hotter nitrous oxide-acetylene flame or the addition of a releasing agent such as lanthanum can help reduce the interference.

Flame atomization produces ions as well as atoms. Since only atoms are detected, it is important that the ratio of atoms to ions remain constant for the element being analyzed. This ratio is affected by the presence of other elements in the sample matrix. The addition of large amounts of an easily ionized element such as potassium to both the sample and standards helps mask the ionization interference.

The capabilities of Flame AA can be extended by employing the following modifications:

1. A cold quartz tube for containing mercury vapor (for mercury determination).

2. A heated quartz tube for decomposing metallic hydride vapors for As, Se, Sb, Pb, Te, Sn, and Bi determination.

3. A graphite quartz tube for decomposing involatile compounds of metals, with extremely high sensitivity.

INDUCTIVELY COUPLED PLASMA
ATOMIC EMISSION SPECTROSCOPY

In inductively coupled plasma atomic emission spectroscopy (ICP), the sample is vaporized and the element of interest atomized in an extremely high temperature ($\sim$ 7000°C) argon plasma, generated and maintained by radio frequency coupling. The atoms collide with energetically excited argon species and emit characteristic atomic and ionic spectra that are detected with a photomultiplier tube. Separation of spectral lines can be accomplished in two ways. In a sequential or scanning ICP (Figure 2), a scanning monochromator with a movable grating is used to being the light from the wavelength of interest to a single detector. In a simultaneous or direct reader ICP (Figure 3), a polychromator with a diffraction grating is used to disperse the light into its component wavelength. Detectors for the elements of interest are set by the vendor during manufacture. Occasionally a scanning channel is added to a direct reader to allow measurement of an element not included in the main polychromator.

ICP is used for the determination of ppm levels of metals in liquid samples. It is not suitable for the noble gases, halogens, or light elements such as H, C, N, and 0. Sulfur requires a vacuum monochromator. A direct reader ICP excels at the rapid analysis of multi-element samples.

Common sample types analyzed by ICP include trace elements in polymers, wear metals in oils, and numerous one-of-a-kind catalysts.

ICP instruments are limited to the analysis of liquids only. Solid samples require some sort of dissolution procedure prior to analysis. The final volume of solution should be at least 25 mL. The solvent can be either water, usually containing 10% acid, or a suitable organic solvent such as xylene.

ICP offers good detection limits and a wide linear range for most elements. With a direct reading instrument multi-element analysis is extremely fast. Chemical and ionization interferences frequently found in atomic absorption spectroscopy are suppressed in ICP analysis. Since all samples are converted to simple aqueous or organic matrices prior to analysis, the need for standards matched to the matrix of the original sample is eliminated.

The requirement that the sample presented to the instrument must be a solution necessitates extensive sample preparation facilities and methods. More than one sample preparation method may be necessary per sample depending on the range of elements requested. Spectral interferences can complicate the determination of trace elements in the presence of other major metals. ICP instruments are not rugged.

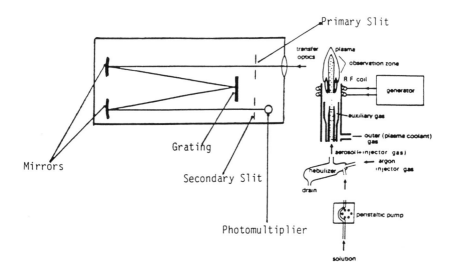

Figure 2. Shows a scanning ICP.

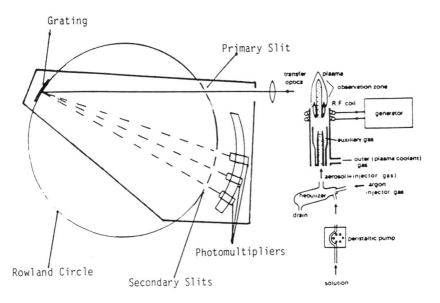

Figure 3. Shows a direct reader ICP.

Constant attention by a trained operator, especially to the sample introduction and torch systems, is essential.

Spectral interferences, such as line overlaps, are prevalent and must be corrected for accurate quantitative analysis.   With a scanning instrument it may be possible to move to an interference free line. With a direct reader, sophisticated computer programs apply mathematical corrections based on factors previously determined on multi-element standards.

ION CHROMATOGRAPHY (IC)

Commercial ion chromatograph instruments have become available since early 1976. Ion chromatography (IC) is a combination of ion exchange chromatography, eluent suppression and conductimetric detection. For anion analysis, a low capacity anion exchange resin is used in the separator column and a strong cation exchange resin in the H+ form is used in the suppressor column. A dilute mixture of Na_2CO_3 $NaHCO_3$ is used as the eluent, because carbonate and bicarbonate are conveniently neutralized to low conductivity species and the different combinations of carbonate-bicarbonate give variable buffered pH values. This allows the ions of interest in a large range of affinity to be separated. The anions are eluted through the separating column in the background of carbonate-bicarbonate and conveniently detected based on electrical conductivity. The reactions taking place on these two columns are, for an anion X:

A. Separator Column

$$Resin - N^+HCO_3^- + NaX^+ \rightleftarrows Resin - N^+X^- + Na^+HCO_3^-$$

B. Suppressor

$$Resin - SO_3^- + Na^+HCO_3^- \rightleftarrows Resin - SO_3^-Na^+ + H_2CO_3$$

$$Resin - SO_3^-H^+ + Na^+X^- \rightleftarrows Resin - SO_3^-Na^+ + H^+X^-$$

As a result of these reactions in the suppressor column, the sample ions are presented to the conductivity detector as H^+X^-, not in the highly conducting background of carbonate-bicarbonate, but in the low conducting background of H_2CO_3.

Figure 4 shows a schematic representation of the ion chromatography system. Dilute aqueous sample is injected at the head of the separator column. The anion exchange resin selectively causes the various sample anions of different types to migrate through the bed at different respective rates, thus effecting the separation. The effluent from the separator column then passes to the suppressor column where the H^+ form cation exchange resin absorbs the cations in the eluent stream. Finally, the suppressor column effluent passes through a conductivity cell. The highly conductive anions in a low background conductance of H_2CO_3 are detected at high sensitivity by the conductivity detector. The nonspecific nature of the conductimetric detection allows several ions to be sequentially determined in the same sample. The conductimetric detection is highly specific and relatively free from interferences. Different stable valance states of the same element can be determined.

On the other hand, because of the nonspecific nature of the conductivity detector, the chromatograph peaks are identified only by their retention times. Thus, the two ions having the same or close retention times will be detected as one broad peak giving erroneous results.

Figure 5 shows a typical chromatogram for the standard common anions F^-, Cl^-, NO_2^-, PO_4^{-3}, Br^-, NO_3 and SO_4^{-2}. Numerous applications of ion chromatography have been illustrated in the literature for a variety of complex matrices.

The advantages of ion chromatography are:

A. Sequential multi-anion capability; eliminates individual determinations of anion by diverse technique.

B. Small sample size (< 1 mL).

C. Rapid analysis (~ 10 minutes for ~ 7 anions).

D. Large dynamic range over four decades of concentration.

E. Speciation can be determined.

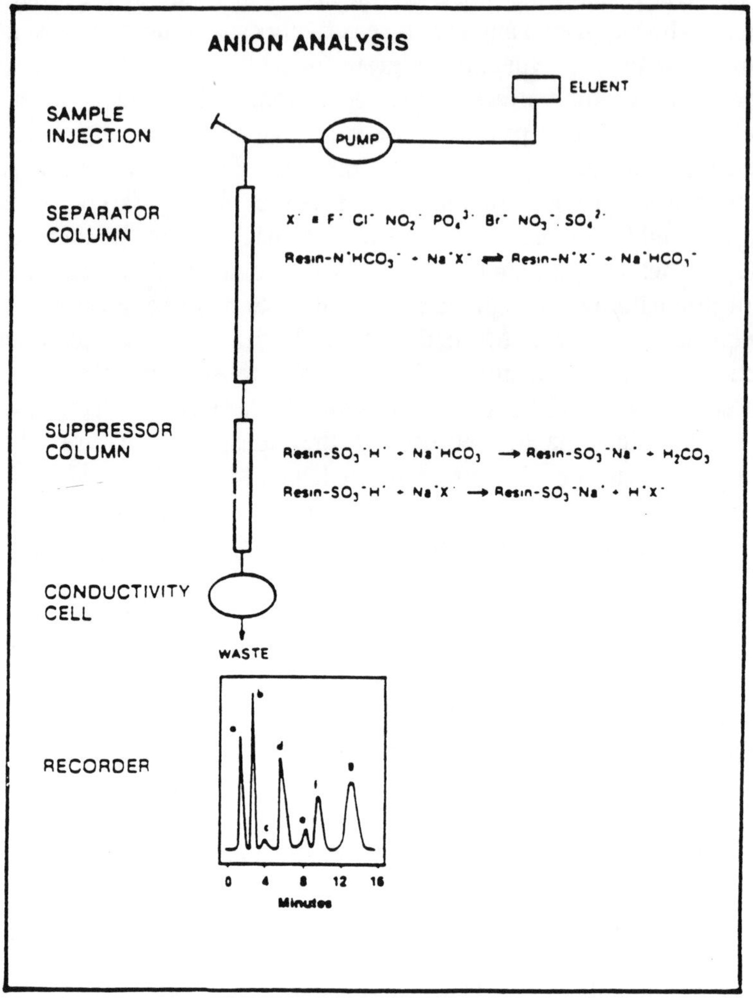

Figure 4. ION Chromatography (IC) flow scheme.

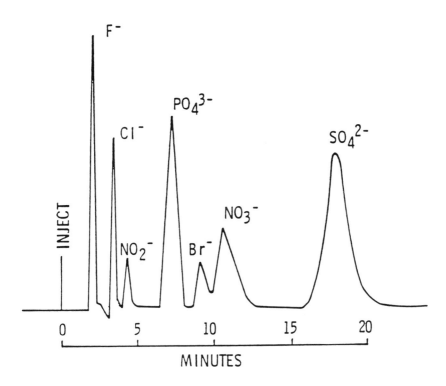

Figure 5.   Shows analysis of standard inorganic anions by ion chromatography.

The principle disadvantages of IC are:

A. Interferences possible if two anions have similar retention times.

B. Determination difficult in the presence of an ion present in very large excess over others.

C. Sample has to be in aqueous solution.

D. Method not suitable for anions with PKa of < 7.

In addition to the common inorganic anions analyzed by IC, a number of other species can also be determined by using appropriate accessories. Some of these applications include:

Technique	Species
A. Ion chromatography	Carboxylic Acids
B. Chemistry - IC	Formaldehyde; Borate
C. Mobile Phase IC	Ammonia; Fatty Acids; Ethanol-Amines
D. Electrochemical Detection	Phenols, CN^-, Br^-, I^-, S^{-2}, etc.

ION SELECTIVE ELECTRODES (ISE)

ISE measures the ion activities or the thermodynamically effective free ion concentrations. ISE has a membrane construction that serves to block the interfering ions and only permit the passage of ions for which it was designed. However, this rejection is not perfect, and hence some interferences from other ions occur. The electrode calibration curves are good over 4 to 6 decades of concentration. The typical time per analysis is about a minute, though some electrodes need 15 minutes for adequate response. The response time is faster as more concentrated solutions are analyzed. Although a single element technique, many elements can be determined sequentially by changing electrodes, provided calibration curves are prepared for all ions. Also, the instrument is portable and is thus useful for field studies. Sample volumes needed are typically about 5 mL, although 300 μL or less can be measured with special modifications. An accuracy of 2-5% is achieved. The ISE measures the activity of the ions in solution. This activity is related to concentration and thus, in effect, measures the concentration. However, if an ion such as fluoride, which complexes with some metals-Fe or Al- is to be measured, it must be decomplexed from these cations by the addition of a reagent such as citric acid or EDTA. ISEs for at least 22 ionic species are commercially available.

An example is described here for the measurement of fluoride ions in solution. The fluoride electrode uses a LaF_3 single crystal membrane and an internal reference, bonded into an epoxy body. The crystal is an ionic conductor in which only fluoride ions are mobile. When the membrane is in contact with a fluoride solution, an electrode potential develops across the membrane. This potential, which depends on the level of free fluoride ions in solution, is measured against an external constant reference potential with a digital pH/mv meter or specific ion meter. The measured potential corresponding to the level of fluoride ions in solution is described by the Nernst equation:

$$E = E_o - S \log A$$

where:

E = measured electrode potential
E_o = reference potential (a constant)
A = fluoride level in solution
S = electrode slope

The level of fluoride, A, is the activity or "effective concentration" of free fluoride ions in solution. The total fluoride concentration, C, may include some bound or complexed ions as well as free ions. The electrode responds only to the free ions, whose concentration is:

$$C_f = C_t - C_b$$

where C_b is the concentration of fluoride ions in all bound or complexed forms.

The fluoride activity is related to free fluoride concentration by the activity coefficient r:

$$A = rC_f$$

Ionic activity coefficients are variable and largely depend on total ionic strength. Ionic strength is defined as:

$$\text{Ionic Strength} = 1/2 \ \Sigma \ C_i Z_i^2$$

where:

$$C_i = \text{concentration of ion i}$$
$$Z_i = \text{charge of ion i}$$

If the background ionic strength is high and constant relative to the sensed ion concentration, the activity coefficient is constant and activity is directly proportional to concentration. Since the electrode potentials are affected by temperature changes, the sample and standard solutions should be close to the same temperature. At the 20 ppm level a 1°C change in temperature gives a 2% error. The slope of the fluoride electrode response also changes with temperature. The electrode can be used at temperatures from 0°C to 100°C, provided that the temperature has equilibrated, which may take as long as an hour. In general, it is best to operate near room temperature.

ISEs are subject to two types of interferences: method interference and electrode interference. In the first type, some property of the sample prevents the electrode from sensing the ion of interest; e.g., in acid solution fluoride forms complexes with H^+ and the fluoride. ISE cannot detect the masked fluoride ions. In the electrode interference, the electrode responds to ions in solution other than the one being measured; e.g., bromide ion poses severe interference in using chloride ISE. The extent of interference depends on the relative concentration of analyze to interfering ions. The interfering ions can be complexed by changing pH or adding a reagent to precipitate them. However, finding the right chemistry is not always easy.

Going back to the example of fluoride determination, the fluoride forms complexes with aluminum, silicon, iron, and other polyvalent cations as well as hydrogen. These complexes must be destroyed in order to measure total fluoride, since the electrode will not detect complexed fluoride. This is achieved by adding a total ionic strength adjustment buffer which contains the reagent CDTA (cyclohexylene dinitrilo tetraacetic acid) which preferentially complex the cations and releases the fluoride ions. The carbonate and bicarbonate anions interfere by making the electrode response slow, hence these ions are eliminated by heating the solution with acid until all CO_2 is removed. At a pH above 7, hydroxyl ions interfere, while at a pH less than 5, the H^+ ions form complexes such as HF_2 thus producing low fluoride results. Addition of TISAB to both samples and standards and further adjustment of pH to between 5.0 and 5.5 is necessary to eliminate the hydroxide interference and the formation of hydrogen-fluoride complexes. Other common anions such as other halides, sulfate, nitrate, phosphate, or acetate do not interfere in the fluoride measurement.

The advantages of this technique are:

A. Inexpensive and simple to use instrument.

B. Rapid analysis; about a minute/sample.

C. Portable instrument; can be used in the field.

D. Large dynamic range over 4-6 decades of concentration.

MASS SPECTROMETRY (MS and GC/MS)

Most of the spectroscopic and physical methods employed by the chemist in structure determination are concerned only with the physics of molecules, mass spectroscopy deals with both the chemistry and the physics of molecules, particularly with gaseous ions. In conventional mass spectrometry, the ions of interest are positively charged ions. The mass spectrometer has three functions:

1. To produce ions from the molecules under investigation.

2. To separate these ions according to their mass to charge ratio.

3. To measure the relative abundances of each ion.

In the 1950s, Benyon, Biemann and McLafferty clearly demonstrated the chemistry of functional groups in directing fragmentation, and the power of mass spectrometry for organic structure determination began to develop.

Today, mass spectrometry has achieved status as one of the primary spectroscopic methods to which a chemist faced with a structural problem turns. The great advantage of the method is found in the extensive structural information which can be obtained from sub-microgram quantities of material.

The methodology of mass separation is governed by both the kinetic energy of the ion and the ion's trajectory in an electromagnetic field. There exists a balance between the centripetal and centrifugal forces which the ion experiences. Centripetal forces are caused by the kinetic energy and centrifugal forces by the electromagnetic field. We may express this force balance as follows, Figure 6 or Figure 7 with GC:

$$\frac{mU^2}{r} = qUB$$

where:

$$
\begin{aligned}
m &= \text{ion's mass} \\
U &= \text{ion's velocity} \\
r &= \text{radius of ion trajectory in the magnetic field} \\
q &= \text{ion's charge} \\
B &= \text{magnetic field strength}
\end{aligned}
$$

The right-hand side is the centripetal force, and the left-hand side is the centrifugal force.

Solving for mass-to-charge ratio yields:

$$\frac{m}{q} = \frac{Br}{U}$$

The kinetic energy of the ion is given by:

$$aV = 1/2\ mU^2$$

where:

$$
\begin{aligned}
q &= \text{charge of the ion} \\
V &= \text{accelerating potential} \\
m &= \text{ion's mass} \\
U &= \text{ion's velocity}
\end{aligned}
$$

Solving for U yields:

$$U = \frac{2qV^{1/2}}{m}$$

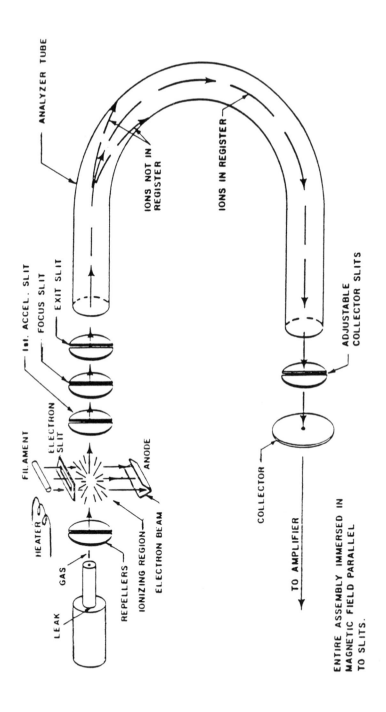

Figure 6. 21-104 Mass spectrometer analyzer.

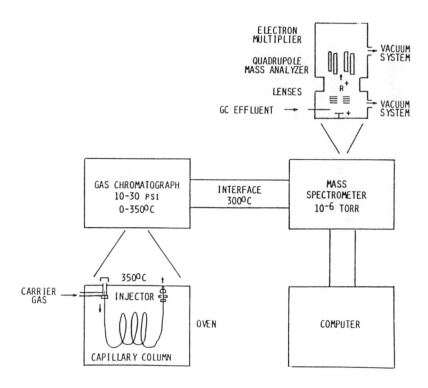

Figure 7. Schematic of a GC/MS instrument.

By substituting for U in the second expression, we obtain

$$m/q = Br/(2qV^{1/2}/m)$$

Squaring each side of the equation yields:

$$\frac{m}{q} = \frac{B^2e^2}{2V}$$

Thus, the mass to charge ratio can be determined if one knows B, e, and V. Since e is constant for a mass passing through the two slivers, scanning a spectrum is achieved by varying either B or V, keeping the other constant.

Mass spectrometers provide a wealth of information concerning the structure of organic compounds, their elemental composition and compound types in complex mixtures. A detailed interpretation of the mass spectrum frequently allows the positions of the functional groups to be determined. Moreover, mass spectrometry is used to investigate reaction mechanisms, kinetics, and is also used in tracer work.

The mass spectrum may be either in analog form (chart paper) or digital form (printed paper). Analyses are calculated to give mole %, weight %, or volume %. Either individual components, compound types by carbon number, or total compound type are reported. This is determined by the nature of the sample and the requirements of the submitter.

The characteristics of the sample submitted for an MS test are:

A. Size: 1-1000 mg.

B. State: Gas, liquid, or solid, but only the portion vaporizable at about 300°C is analyzed.

C. Phases: If sample has more than one phase, each phase can generally be analyzed separately.

D. Composition Limitations: Essentially no limits to composition, simple mixtures, and complex mixtures can be handled.

E. Temperature Range: Samples should be at room temperature and should be thermally stable up to 300°C for bath introduction and may be involatile for field desorption work.

A wide variety of materials from gases to solids and from simple to complex mixtures can be analyzed. The molecular weight and atomic composition are generally determined. Only a very small amount of sample is required. Most calibration coefficients can be used for long periods of time.

Some compounds such as long chain esters and polyethers decompose in the inlet system, and the spectrum obtained is not that of the initial substance. Calibration coefficients are required for quantitative analyses. The sample introduced to the instrument cannot usually be recovered.

Some classes of compounds, such as olefins and naphthenes, give very similar spectra and cannot be distinguished except by analysis before and after hydrogenation or dehydrogenation.

NUCLEAR MAGNETIC RESONANCE SPECTROMETER

Nuclear Magnetic Resonance (NMR) is a spectrometric technique for determining chemical structures. When an atomic nucleus with a magnetic moment is placed in a magnetic field, it tends to align with the applied field. The energy required to reverse this alignment depends on the strength of the magnetic field and to a minor extent on the environment of the nucleus, i.e., the nature of the chemical bonds between the atom of interest and its immediate vicinity in the molecule. This reversal is a resonant process and occurs only under select conditions. By determining the energy levels of transition for all of the atoms in a molecule, it is possible to determine many important features of its structure. The energy levels can be expressed in terms of frequency of electromagnetic radiation, and typically fall in the range of 5-600 MHz for high magnetic fields. The minor spectral shifts due to chemical environment are the essential features for interpreting structure and are normally expressed in terms of part-per-million shifts from the reference frequency of a standard such as tetramethyl silane.

The most common nuclei examined by NMR are ^{1}H and ^{13}C, as these are the NMR sensitive nuclei of the most abundant elements in organic materials. ^{1}H represents over 99% of all hydrogen atoms, while ^{13}C is only just over 1% of all carbon atoms; further, ^{1}H is much more sensitive than $1 \sim C$ on an equal nuclei basis. Until fairly recently, instruments did not have sufficient sensitivity for routine ^{13}C NMR, and ^{1}H was the only practical technique. Most of the time it is solutions that are characterized by NMR, although ^{13}C NMR is possible for some solids, but at substantially lower resolution than for solutions.

In general, the resonant frequencies can be used to determine molecular structures. ^{1}H resonances are fairly specific for the types of carbon they are attached to, and to a lesser extent to the adjacent carbons. These resonances may be split into multiples, as hydrogen nuclei can couple to other nearby hydrogen nuclei. The magnitude of the splittings, and the multiplicity, can be used to better determine the chemical structure in the vicinity of a given hydrogen. When all of the resonances observed are similarly analyzed, it is possible to determine the structure of the molecule. However, as only hydrogen is observed, any skeletal feature without an attached hydrogen can only be inferred. Complications can arise if the molecule is very complex, because then the resonances can overlap severely and become difficult or impossible to resolve.

^{13}C resonances can be used to directly determine the skeleton of an organic molecule. The resonance lines are narrow and the chemical shift range (in ppm) is much larger than for ^1H resonances. Furthermore, the shift is dependent on the structure of the molecule for up to three bonds in all directions from the site of interest. Therefore, each shift becomes quite specific, and the structure can be easily assigned, frequently without any ambiguity, even for complex molecules.

Very commonly, however, the sample of interest is not a pure compound, but is a complex mixture such as a coal liquid. As a result, a specific structure determination for each molecular type is not practical, although it is possible to determine an average chemical structure. Features which may be determined include the hydrogen distribution between saturate, benzylic, olefinic, and aromatic sites. The carbon distribution is usually split into saturate, heterosubstituted saturate, aromatic + olefinic, carboxyl, and carbonyl types. More details are possible, but depend greatly on the nature of the sample, and what information is desired.

Any gas liquid or solid sample that can be dissolved in solvents, such as CCl_4, $CH \sim Cl$, acetone or DMSO to the one percent level or greater can be analyzed by this technique. Samples of $\sim 0.1g$ or larger of pure material are sufficient. Solids can also be analyzed as solid. However, special arrangements need to be made. In either case, the analysis is non-destructive so that samples can be recovered for further analysis if necessary.

The NMR experiment can be conducted in a temperature range from liquid nitrogen (-209°C) to +150°C. This gives the experimenter the ability to slow down rapid molecular motions to observable rates or to speed up very slow or viscous motions to measurable rates.

NMR is a very powerful tool. It often provides the best characterization of compound structure, and may provide absolute identification of specific isomers in simple mixtures. It may also provide a general characterization by functional groups which cannot be obtained by any other technique. As is typical with many spectroscopic methods, adding data from other techniques (such as mass or infrared spectrometry) can often provide greatly improved characterizations.

The following are general notes and comments concerning the use of NMR specifically for common rubber characterization problems. A schematic of a Fourier Transform NMR spectrometer is given in Figure 8.

Sample Preparation

Samples are analyzed by proton (^{1}H) and/or Carbon ^{13}C NMR.

Sample requirements   ~ 1/2 gm for ^{1}H
                       ~ 1 gm for ^{13}C

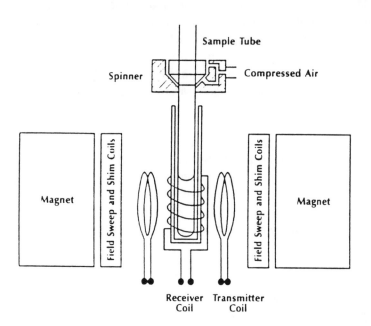

Block diagram of a high-resolution NMR spectrometer and the arrangement of the sample in the probe (cross coil configuration).

Figure 8. Schematic of a Fourier Transform NMR spectrometer.

For butyl based polymers dissolving is performed in dewatered chloroform at ambient temperature. EP (ethylene-propylene) can be dissolved in deuterated O-Dichlorobenzene at 140°C.

Wet samples cannot be accurately analyzed. Opaqueness in a rubber sample is generally an indication of moisture in the sample.

Established Methods by ^{1}H NMR

Sample Type	Analyzed for
EPDM	ENB, Hexdiene, Ethylene, DCDP (Wt. %)
SBB	Isoprene Unsat. (Mole %); KR01 (Wt. %)

Sample Type (continued)	Analyzed for
Butyl Bromobutyl Chlorobutyl	Isoprene Unsat. (Mole %) Type I, II, III (%)

I	II	III
CH$_3$	CH$_2$	CH$_2$X
$-$ CH$_2$ - C = CH - CH$_2$ $-$	$-$ CH$_2$ - C - CH - CH$_2$ $-$ (with X below C)	$-$ CH$_2$ - C = CH - CH$_2$ $-$

Butyl   Isoprene Unsat. (Mole %)

CH$_3$
|
$-$ CH$_2$ - C = CH - CH$_2$ $-$ 1,4 Isoprene

CH$_3$
|
$-$ CH$_2$ - C $-$ 1,2 Isoprene
|
CH
|
CH$_2$

Where X = Cl or Br.

NMR results are quantitative. Analysis of a ^{13}C or ^{1}H spectrum would reveal the different types of functionalities, as well as their contents in the sample. For example, Figure 9 shows the ^{1}H NMR spectrum of the diene (ENB) in an EPDM polymer (ethylene-propylene diene monomer).

Figure 10 shows a ^{13}C NMR spectrum of EP used to determine the sequence distribution. The detectability for ^{1}H NMR is typically 0.01 mole % and for ^{13}C NMR is 0.1 mole %.

FOURIER TRANSFORM INFRARED (FTIR)

Generally, this technique is used to analyze samples that are available either in small quantity or a small entity. Gels within a rubber sample, have to be microtomed (i.e., cut into very thin slices) and mounted in KBr plates in a Microscopy Lab. Samples contaminated with inorganic components are usually analyzed by both X-ray and FTIR Microscope. Sample size ~ 20 microns can be analyzed by the FTIR-microscope.

FTIR (refer to Figure 11) is commonly used for qualitative identification of various functionalities. For quantitative analysis, FTIR requires the use of well characterized standards. NMR spectroscopy is typically used to characterize a set of samples which are then used as standards for the FTIR calibration.

Figure 12 shows the FTIR spectrum of a butyl sample. This sample contains BHT, ESBO, stearic acid, and calcium stearate. The contents of all these components can be determined from this single spectrum. In some cases, the assigned peak absorbance is relatively small, so a thick film, ~ .6 mm, is used.

ULTRAVIOLET, VISIBLE, AND
INFRARED SPECTROMETRY (UV, Vis, IR)

When electromagnetic radiation passes through a sample, some wavelengths are absorbed by the molecules of the sample. Energy is transferred from the radiation to the sample, and the molecules of the sample are said to be elevated to an excited energy state. The total energy state of the ensemble of molecules may be regarded as the sum of the four kinds of energy: electronic, vibrational, rotational, and

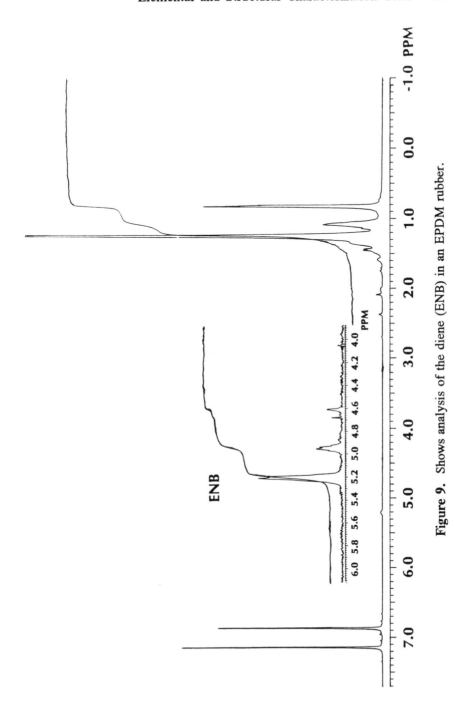

Figure 9. Shows analysis of the diene (ENB) in an EPDM rubber.

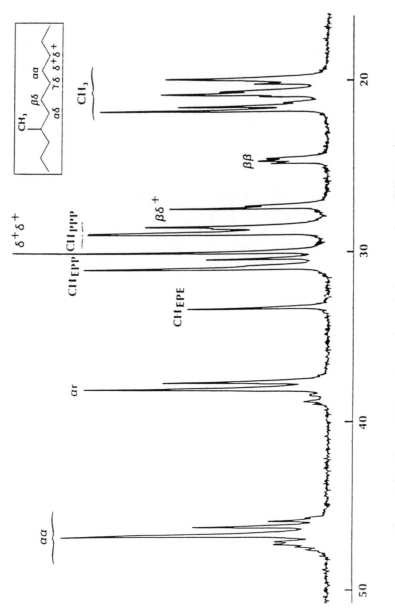

Figure 10. Shows 13C spectrum of an ethylene-propylene (EP) copolymer.

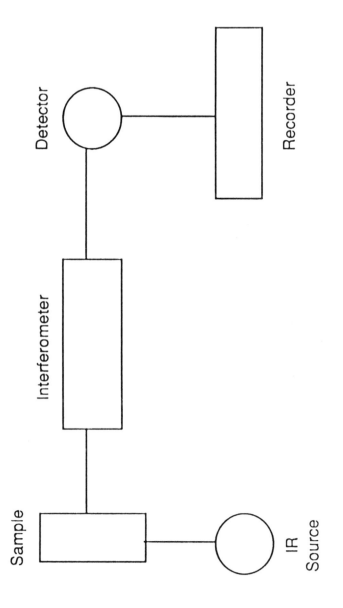

Figure 11. Shows the basic components of an FTIR.

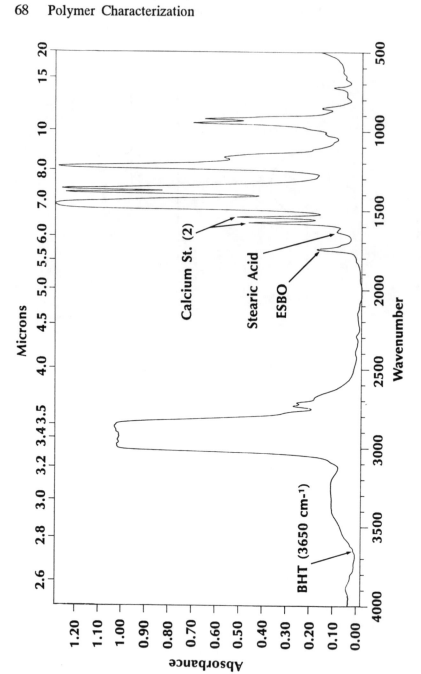

Figure 12. Example of an FTIR spectrum for butyl rubber.

transnational. Transnational energy is associated with an elevation of the temperature of the sample. Rotational energy comes about by the absorption of very high wavelengths of infrared radiation (25-500μm), and is manifested by an increase in the rotational energy of the sample molecules. Vibrational energy arises when radiation in the mid-infrared region is absorbed (2-25μm), and is manifested by an increase in the vibrational energies of functional groups within the sample molecule. Electronic energy is gained by an ensemble of molecules when an electron is promoted to a higher molecular orbital by absorption in the ultraviolet and visible regions of the spectrum (0.2-0.8μm).

Pure transitions between rotational states represent very small energy changes (high wavelength). Absorption spectra observed in the far infrared are generally "pure" in the sense that the energy absorbed by the molecule is entirely converted into pure rotational motion. This is not the case in the other regions of the spectrum. Thus, when higher amounts of energy are absorbed by molecules, the vibrational motions generated are not restricted to those for which the rotational properties of the molecule remain constant. The absorption band, therefore, will represent a composite of vibrational motions, each occurring in molecules of different rotational levels. The same is true for electronic absorptions, where both rotational and vibrational properties of the molecules are impressed on the electronic transitions.

Another complication arises in the interpretation of absorption spectra. If a molecule vibrates with pure harmonic motion and the dipole moment is a linear function of the displacement, then the absorption spectrum will consist of fundamental transitions only. If either of these conditions is not met, as is usually the case, the spectrum will contain overtones (multiples of the fundamental) and combination bands (sums and differences). Most of these overtones and combination bands occur in the near-infrared (0.8-2.0μm).

Not all vibrations and rotations are infrared-active. If there is no change in dipole moment, then there is no oscillating electric field in the motion, and there is no mechanism by which absorption of electromagnetic radiation can take place. An oscillation, or vibration, about a center of symmetry, therefore, will not be observed in the infrared spectrum (absorption) but can be observed in the Raman spectrum (scattering).

In summary, therefore, there are five regions of the electromagnetic spectrum of interest:

<u>m</u>

0.2-0.4	Ultraviolet (electronic)
0.4-0.8	Visible (electronic)
0.8-2.0	Near-IR (overtones)
2.0-25.0	Mid-IR (vibrational)
25.0-500.0	Far-IR (rotational)

Electronic transitions (UV, visible spectra) generally give information about unsaturated groups in the sample molecules. Olefins absorb near $0.22\mu m$, aromatics near $0.26\text{-}0.28\mu m$, carbonyls near $0.20\text{-}0.27\mu m$, poly-nuclear aromatics near $0.26\text{-}0.50\mu m$, and conjugated $C=S$ groups near $0.62\mu m$. Any material which is colored will generally show absorption in the visible region. The intensity of the absorption is proportional to the number of chromophores giving rise to the absorption band.

Overtones (near-IR) are useful for studying the presence of groups containing hydrogen. Fundamentals involving hydrogen vibration tend to congregate near the same frequencies in the mid-IR, but are easier to distinguish and study in the overtone region.

Vibrational transitions (mid-IR) are the most useful of all to study. These give information about the presence or absence of specific functional groups in a sample. Practically all functional groups (that have an infrared-active fundamental) display that fundamental over a very narrow range of wavelength in the mid-infrared region. Moreover, the whole spectrum, containing fundamentals, overtones, and combination bands, constitutes a fingerprint of the sample. This means that although we might not know what a sample is, we will always know it later if it occurs again. Finally, the absorption intensity of any band, whether fundamental or overtone, is proportional to the number of functional groups giving rise to the signal.

The characteristics of the sample used are as follows:

A. For gases, we generally need about 250 cm³ at 1 atm to obtain a spectrum.

B. For liquids, we generally need about 0.25 cm³ to obtain a spectrum.

C. For solids, we generally need about 1 mg to obtain a spectrum.

D. Trace analyses within samples will, of course, increase the sample requirements proportionally.

The advantages of this technique are:

A. Faster and cheaper than most other techniques.

B. Very specific for certain functional groups.

C. Very sensitive for certain functional groups.

D. Fingerprint capability.

The disadvantages of this technique are:

A. Requires special cells, NaCl, KBr, quartz, etc.

B. Usually requires solubility of sample.

C. Very difficult to get good quantitation in solids.

D. Must calibrate all signals.

E. Water interferes.

Measurement interferences can occur from:

A. Water interferes with practically all IR work.

B. Solvents generally interfere and must be selected carefully.

C. Multicomponent samples generally have mutually interfering species. Separations are often required. Sometimes, changing the spectral region helps.

D. Optical components interfere to different extents in different regions. Thus, quartz is good for UV/Vis/Near-IR, but bad for mid-IR/far-IR. KBr is good for mid-IR, bad for far-IR.

Fourier Transform Infrared Spectrometry is a special technique. In dispersive spectrometry, the wavelength components of light are physically separated in space (dispersed) by a prism of grating (Figure 13). Modern dispersive spectrometers divide the incident beam into two beams; one beam goes through the sample and the other goes through a suitable reference material. The intensity of both beams are monitored by a suitable detector, and final data output can be displayed in either transmittance or absorbance:

$$transmittance \; = \; I_s/I_R$$

$$absorbance \; = \; - \log I_s/I_R$$

where I_s, I_R refer to the intensities in the sample beam and reference beam, respectively. This rationing occurs at each wavelength element, and the final plot is a graphical display with transmittance or absorbance on the Y axis, and wavelength or frequency on the X axis.

In Fourier transform spectrometry, the wavelength components of light are not physically separated. Instead, the light is analyzed in the time frame of reference (the time domain) by passing it through a Michelson interferometer. The Michelson interferometer is so constructed that light is separated into two beams by a beamsplitter. One beam strikes a stationary mirror and is reflected back to the beamsplitter.

The other beam strikes a moving mirror and is reflected back to the beamsplitter. The two beams are recombined at the beamsplitter and proceed on to the sample and the detector. Note that, upon combination, the two beams will interfere constructively or destructively, depending upon whether the difference in pathlength of the two beams is an integral multiple of the wavelength. The difference in pathlength is called the retardation, and when the retardation is an integral multiple of the wavelength, the interference is maximally constructive. If we plot retardation vs. intensity measured at the detector, we have, in effect, a time domain function of the intensity, and this can be transformed by Fourier transform mathematical techniques into a frequency function of the intensity.

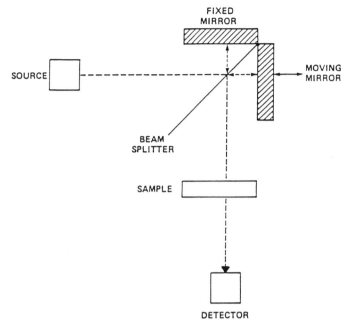

Figure 13. Shows the Michelson interferometer.

Ultimately, then, we get the same information by both techniques, but we get it much faster, and more precisely, by the Fourier transform technique.

X-RAY FLUORESCENCE SPECTROMETRY

X-ray fluorescence spectrometry (XRF) is a non-destructive method of elemental analysis. XRF is based on the principle that each element emits its own characteristic X-ray line spectrum. When an X-ray beam impinges on a target element, orbital electrons are ejected. The resulting vacancies or holes in the inner shells are filled by outer shell electrons. During this process, energy is released in the form of secondary X-rays known as fluorescence. The energy of the emitted X-ray photon is dependent upon the distribution of electrons in the excited atom. Since every element has a unique electron distribution, every element produces a unique secondary X-ray spectrum, whose intensity is proportional to the concentration of the element in the sample. The excitation process and resulting X-ray spectrum are illustrated for calcium in Figure 14.

X-ray fluorescence instrumentation is divided into two types, wavelength dispersive (WDXRF) and energy dispersive (EDXRF) spectrometer, both of which are often automated with extensive computer systems for unattended operations that include data collection, reduction, and presentation. In a wavelength dispersive spectrometer (Figure 15), radiation emitted from the sample impinges on an analyzing crystal. The crystal diffracts the radiation according to Bragg's Law and passes it on to a detector which is positioned to collect a particular X-ray wavelength. Most spectrometers have two detectors and up to six crystals to allow optimization of instrument conditions for each element.

In an energy dispersive spectrometer (Figure 16), the emitted X-ray radiation from the sample impinges directly on a solid state lithium drifted silicon detector. This detector is capable of collecting and resolving a range of X-ray energies at one time. Therefore, the elements in the entire periodic table or in a selected portion can be analyzed simultaneously. Optimization for specific elements is accomplished through the use of secondary targets and/or filters. In some cases, radioisotope sources are used in place of X-ray tubes in instruments designed for limited element applications.

XRF offers a unique approach for rapid, non-destructive elemental analysis of liquids, powders, and solids. Although the first row transition elements are the most sensitive, elements from atomic number 12 (magnesium) and greater can be measured over a dynamic range from trace (ppm) to major (percent) element concentrations. EDXRF is well suited for qualitative elemental identification of unique samples, while WDXRF excels at high precision quantitative analysis.

Whenever quantitative analysis is desired, care must be taken to use proper standards and account for interelement matrix effects since the inherent sensitivity of the method varies greatly between elements. Methods to account for matrix effects include standard addition, internal standard and matrix dilution techniques as well as numerous mathematical correction models. Computer software is also available to provide semi-quantitative analysis of materials for which well-matched standards are not available.

XRF is used by both research and plant laboratories to solve a wide variety of elemental analysis problems. Among the most common are

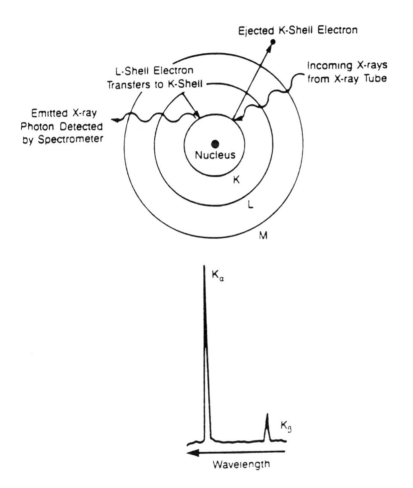

Figure 14.   Illustrates XRF excitation process and resulting spectrum for calcium.

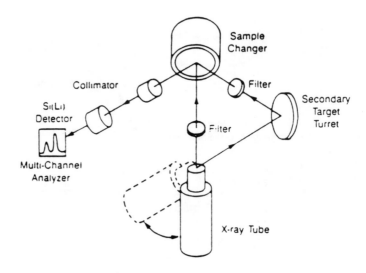

Figure 15. Illustrates the WDXRF spectrometer.

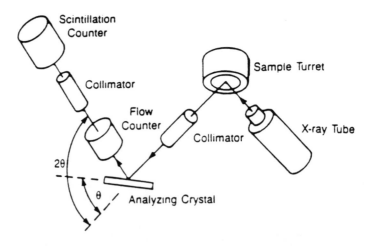

Figure 16. Illustrates the EDXRF spectrometer.

the quantitative analysis of additive elements (Ba, Ca, Zn, P, and S) in additives and lubricating oils, lead, and sulfur in gasoline, sulfur in crudes and fuel oils, and halogens in polymers. The high analytical precision of WDXRF has enabled the development of methods for the precious metal assay of fresh reforming catalyst that rival the precision of classical wet chemical methods. Percent active metal such as Pt, Ir, Re, or Ru as well as trace contaminants such as S and Cl have also been determined on numerous catalyst types.

The elemental composition of unknown materials such as engine deposits can be determined qualitatively and the information used to develop dissolution methods prior to analysis by inductively coupled plasma atomic emission spectroscopy (ICPAES). Alternatively, a semi-quantitative analysis can be provided by XRF alone, especially important when only a limited quantity of sample is available and needed for subsequent tests. The deposit does not even have to be removed from the piston since large objects can be placed directly inside an EDXRF spectrometer.

Aqueous and organic liquids, powders, polymers, papers, and fabricated solids can all be analyzed directly by XRF. The method is nondestructive, so unless dilution is required, the original sample is returned to the submitter. Although the method can be applied to the analysis of materials ranging in size from milligram quantities to bulk parts such as engine pistons, a minimum of 5 grams of sample is usually required for accurate quantitative analysis.

One of the biggest advantages of XRF over other methods of elemental analysis is that it is nondestructive and requires minimal sample preparation. WDXRF, when properly calibrated, offers precision and accuracy comparable to wet chemical methods of analysis. EDXRF offers rapid qualitative analysis of total unknowns.

Due to numerous interelemental matrix effects, matrix matched standards including a blank are necessary for accurate quantitative analysis. The detection limits for XRF are not as low as other spectrometric methods and a noticeable drop-off in sensitivity is noted for light elements such as magnesium.

The most common interferences are absorption and/or enhancement of the element of interest by other elements in the matrix. Line overlaps may also occur. In the analysis of solids, particle size and geological effects can be important. Computer programs are available to correct for all of these interferences.

POTENTIOMETRIC TITRATIONS

A titration is a technique for determining the concentration of a material in solution by measuring the volume of a standard solution that is required to react with the sample. One of the most common titrations is the acid-base titration in which the concentration of a base can be determined by adding a standard solution of an acid to the sample until the base is exactly neutralized. The exact neutralization point is found by the use of an indicator that changes color when the end-point is reached.

There are, however, various cases where the visual method of detecting the end-point cannot be used. For example, the solution may be dark or no appropriate indicator is available. In such cases, physicochemical techniques can be employed. A potentiometric titration is one of this type wherein the end-point is detected by an abrupt change in voltage (between an electrode and the body of the solution) that may be observed as titrant is added to the solution.

As the solution is being titrated, the potential difference that exists between the indicating electrode and the solution may be continuously monitored with a voltage measuring device such as a recorder. This affords an objective means of determining the end-point which occurs when a very slight excess of titrant (as little as 0.25 mL) causes a sharp voltage change. Automatic recorders are now in use which not only plot the titration curve but also electronically determine the end-point and can (if so programmed) complete the calculation producing the finished result.

Results are reported in any of the following units:

Wt %, mg/liter, normality, milliequivalents/100 gm

In some cases a plot of mL of titrant vs. voltage can be provided. Such plots are useful to determine whether there is more than one titratable material in the sample and to learn something about the character of the material titrated.

Titratable functional groups such as chloride, sulfide, mercaptide, weak or strong acids, weak or strong bases, and certain amines may be determined by this technique.

The sample characteristics are as follows:

- *Size*--The amount of sample to be submitted depends strictly upon the concentration of the functional group sought. For samples whose functional group is expected to be in the ppm range, it is advisable to use 100 mL of sample.
- *State*--Sample must be a solid or liquid.
- *Composition limitation*--Sample must be soluble in water or in one of the several special non-aqueous titration solvents available that can accommodate most petroleum fractions (except some of the heavy ends).
- *Temperature*--Titrations are conducted at room temperature.
- *Concentration*--A wide range of concentrations can be accommodated by varying the amount of sample dissolved in the titration solvent. In many cases concentrations as low as a few ppm can be reported.

Titrations are relatively simple and rapid. They provide information concerning chemically reactive functional groups that would be difficult to obtain by other techniques.

Generally, the potentiometric titration technique is not good for qualitative purposes. One must indicate a priori what functional group he wants determined. The voltage at which the end-point occurs does provide a clue to the material being titrated, but the "voltage spectrum" is too compressed to provide identification by voltage only. For example, strong acids can be differentiated from weak acids but the identity of the acid must be ascertained by other chemical or physical means.

The time required to titrate a single sample may vary somewhat depending upon the nature of the sample and what determinations are requested. An average of about one hour is needed to run a single sample; however, the time per sample may be less when a series of similar samples are run consecutively. If special solutions have to be prepared, an additional two hours may be required for the first sample in the series.

NEUTRON ACTIVATION ANALYSIS

Neutron activation analysis is a method of elemental analysis in which nonradioactive elements are converted to radioactive ones by neutron bombardment, and the elements of interest are determined from resulting radioactivity (Figure 17). High energy (14 MeV) neurons are generated by the reaction of medium energy deuterium ions with tritium. For oxygen analysis, the carefully weighed sample is irradiated for 15 seconds to convert a small amount of the oxygen-16 to nitrogen-16, which emits gamma rays with a half life of 7.4 seconds. he irradiated sample is transferred to a scintillation detector where the gamma rays are counted for 30 seconds to insure that all usable radioactivity has been counted and that no significant radioactivity remains in the sample. The system is calibrated with standards of known oxygen content.

The raw data consists of counts per 30 seconds from a digital counter, which can be converted to weight % of the element of interest.

The characteristics of a proper sample are:

- *Size*--Container is plastic cylinder 9 mm I.D. x 20 mm deep. Holds ~ 1. cc. Sample should fill this container for best accuracy.
- *State and phases*--Solid or liquid, reasonably homogeneous.
- *Composition limitations*--Must be moisture-free if true sample oxygen is desired.
- *Temperature range*--Room temperature only.
- *Concentration of oxygen which can be determined*--0.01-60%.

The advantages of this technique are:

A. Principle method for determining total oxygen directly.

B. Fast (about 10 minutes per analysis). Repeat analysis on weighed sample requires only 1 minute).

C. Nondestructive.

D. Moderate sensitivity.

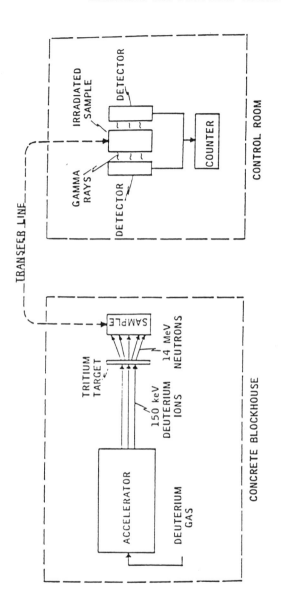

Figure 17. Shows the neutron activation analysis system.

5   RHEOMETRY

Although there are numerous rheometric techniques used, this section will only describe three common systems heavily employed for polymer characterization.

RHEOMETRICS SYSTEM IV

A schematic of the system is illustrated in Figure 1.   For dynamic frequency sweeps (refer to Figure 2), the polymer is strained sinusoidally and the stress is measured as a function of the frequency.   The strain amplitude is kept small enough to evoke only a linear response.   The advantage of this test is that it separates the moduli into an elastic one, the dynamic storage modulus (G') and into a viscous one, the dynamic loss modulus (G").   From these measurements one can determine fundamental properties such as:

- Zero shear viscosity (which can be related to weight average molecular weight and long chain branching).
- Tan delta (which is related to the damping properties).
- Plateau modulus (which can indicate the extent and "tightness" of crosslinking).
- Complex viscosity (which can be related to the steady shear viscosity).

Differences in G' and G" and hence in the properties mentioned above will be found if there are differences in molecular weight, molecular weight distribution (MWD) or long chain branching.   For example, if the MWD is primarily in the high molecular weight end, then the value of G' will be higher.

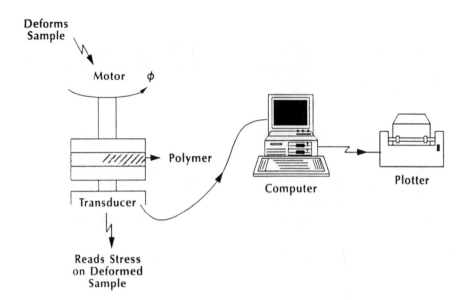

Figure 1. Details of Rheometrics System IV.

To obtain measurements during oscillatory shear, the drive motor causes the fixture to oscillate from high to low shear rates deforming the sample. The transducer detects the periodic stress which is generated by the deformation. The magnitude of the stress is converted into dynamic shear moduli.

The application of stress relaxation is shown in Figure 3. The relaxation modulus (G) is determined after a step strain as a function of time. A step strain is applied to the sample causing a stress. The modulus is measured as the stress relaxes. The stress relaxation modulus shows how molecular weight affects the relaxation process as a function of time as depicted in Figure 4.

To perform a temperature sweep, the sample is deformed at constant frequency over a temperature range. The modulus is then obtained as a function of temperature at constant frequency. This measurement gives the temperature dependence of the rheological properties as well as a good indication of the thermal stability.

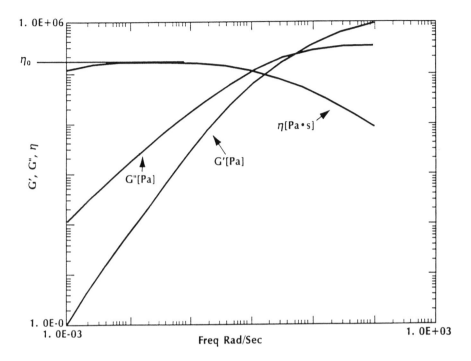

Figure 2. Shows typical frequency sweep.

CAPILLARY RHEOMETRY

Steady shear viscosities can be measured with two different instruments. The System IV can measure polymer viscosities from about 0.001 to 10 sec^{-1} while the Gottfert Capillary Rheometer is capable of obtaining viscosities from 0.1 to 100,000 l/s. In steady shear, the strains are very large as opposed to the dynamic measurements that impose small strains. In the capillary rheometer, the polymer is forced through a capillary die at a continuously faster rate. The resulting stress and viscosity are measured by a transducer mounted adjacent to the die. A schematic of the system is illustrated in Figure 5.

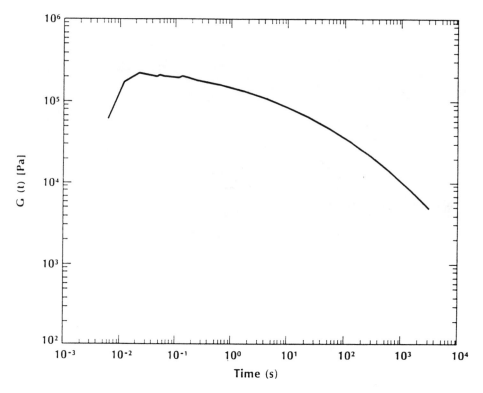

Figure 3. Shows typical stress relaxation.

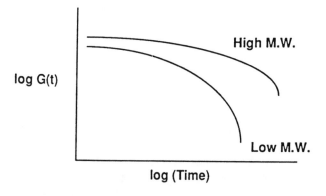

Figure 4.   Shows how the stress relaxation modulus is related to molecular weight.

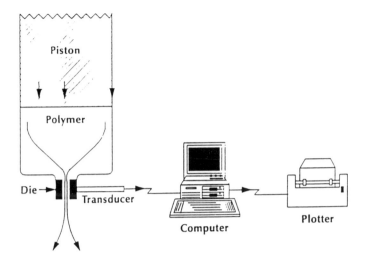

Figure 5. Details of the capillary rheometer.

The shear viscosity can be used for relating the polymer flow properties to the processing behavior, extruder design, and many other high shear rate applications. Elongational viscosity, die swell measurements as well as residence time effects can be estimated. Typical data are shown in Figure 6.

TORQUE RHEOMETRY

Torque rheometers are multipurpose instruments well suited for formulating multicomponent polymer systems, studying flow behavior, thermal sensitivity, shear sensitivity, batch compounding, and so on. The instrument is applicable to thermoplastics, rubber (compounding, cure, scorch tests), thermoset materials, and liquid materials.

When the rheometer is retrofitted with a single-screw extruder one can measure rheological properties and extrusion processing characteristics to differentiate lot-to-lot variance of polymer stocks. It also enables the process engineer to simulate a production line in the laboratory and to develop processing guidelines.

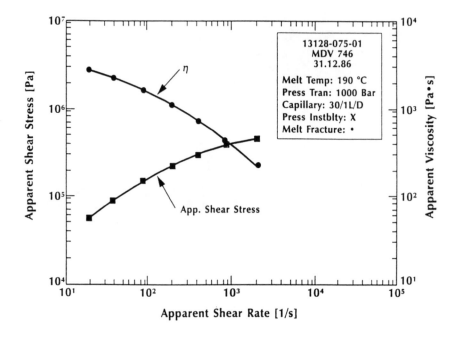

Figure 6. Shows typical capillary rheometer data.

The torque rheometer with a twin-screw extruder is considered a scaled-down continuous compounder. It allows the compounding engineer to develop polymer compounds and alloys. It also permits the formulation engineer to assure that the formulation is optimum.

The torque rheometer is essentially an instrument that measures viscosity-related torque caused by the resistance of the material to the shearing action of the plasticating process.

Torque can be defined as the effectiveness of a force to produce rotation. It is the product of the force and the perpendicular distance from its line of action to the instantaneous center of rotation.

The prevalent design is a microprocessor-controlled torque rheometer. The system consists of two basic units: an electromechanical drive unit and a microprocessor unit. The basic system is shown in Figure 7.

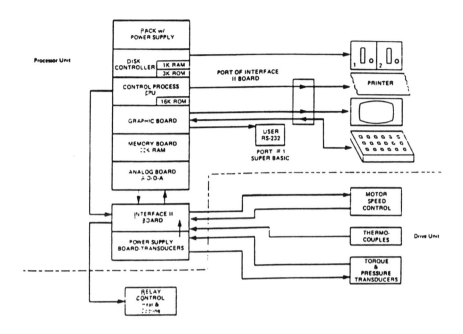

Figure 7.   Schematic diagram of Haake-Buechler System 40 microprocessor.

Most plastics and elastomeric products are not pure materials but rather, mixtures of the basic polymer with a variety of additives, such as pigments, lubricants, stabilizers, antioxidants, flame retardants, antiblock agents, cross-linking agents, fillers, reinforcement agents, plasticizers, UV absorbants, foaming agents, and others. All these additives must be incorporated into the polymer prior to fabrication. Some of the additives take a significant portion of the mixture; others, only minute amounts. Some are compatible; others are not. Depending on the quality of resin and additives and homogenization of the mixtures, the quality of the final product will be varied.   Therefore, developing a quality resin and additives that meet with desired physical and mechanical properties of the product and quality control associated with them play an important role in the plastic industry.   In the development of formulations and applications for new polymers, the torque rheometer is an invaluable instrument. Common practice is to equip the torque rheometer with a miniaturized internal mixer (MIM) to simulate large-scale production at the bench. The mixer generally consists of a mixing chamber shaped like a figure eight with a spiral-lobed rotor in each chamber. A totally enclosed mixing chamber contains two fluted mixing rotors that revolve in opposite directions and at different speeds to achieve a shear action similar to a two-roll mill.

In a chamber, the rotors rotate in order to effect a shearing action on the material mostly by shearing the material repeatedly against the walls of the mixing chamber. This is illustrated conceptually in Figure 8. The rotors have chevrons (helical projections) which perform additional mixing functions by churning the material and moving it back and forth through the mixing chamber. The mixture is fed to the mixing chamber through a vertical chute with a ram. The lower face of the ram is part of the mixing chamber. There is usually a small clearance between the rotors, which rotate at different speeds at the chamber wall. In these clearances, dispersive mixing takes place. The shape of the rotors and the motion of the ram during operation ensure that all particles undergo high intensive shearing flow in the clearances.

There are three sets of interchangeable rotors available on the market. They are roller, cam, and sigma rotors designs, although there are many other designs as illustrated in Figure 9. Normally, roller rotors are used for thermoplastics and thermosets, cam rotors are for rubber and elastomers, and sigma rotors are for liquid materials. Banbury rotors are used with miniaturized Banbury mixer for rubber compounding formulation.

Plastic materials with two or more components being processed should be well mixed so that the compounded material provides the best physical properties for the final product. There are distinctive types of mixing processes. The first involves the spreading of particles over position in space (called distributive mixing). The second type involves shearing and spreading of the available energy of a system between the particles themselves (dispersive mixing). In other words, distributive mixing is used for any operation employed to increase the randomness of the spatial distribution of particles without reducing their sizes. This mixing depends on the flow and the total strain, which is the product of shear rate and residence time or time duration. Therefore, the more random arrangement of the flow pattern, the higher the shear rate; and the longer the residence time, the better the mixing will be.

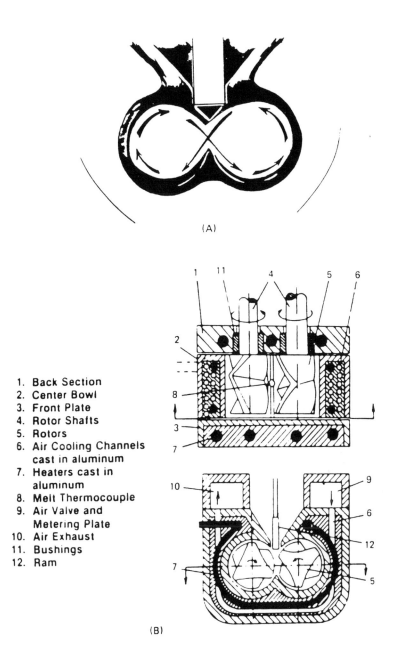

1. Back Section
2. Center Bowl
3. Front Plate
4. Rotor Shafts
5. Rotors
6. Air Cooling Channels
   cast in aluminum
7. Heaters cast in
   aluminum
8. Melt Thermocouple
9. Air Valve and
   Metering Plate
10. Air Exhaust
11. Bushings
12. Ram

Figure 8.   Mixing action in a MIM.

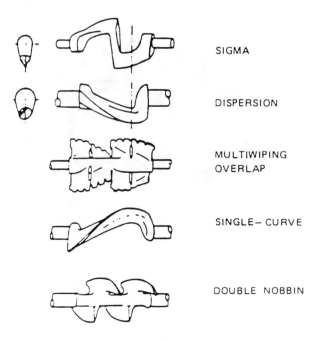

SIGMA

DISPERSION

MULTIWIPING
OVERLAP

SINGLE– CURVE

DOUBLE NOBBIN

Figure 9. Types of mixer rotor configurations.

A dispersive mixing process is similar to that of a simple mixing process, except that the nature and magnitude of forces required to rupture the particles to an ultimate size must be considered. Essential to intensive mixing is the incorporation of pigments, fillers, and other minor components into the matrix polymer. This mixing is a function of shear stress, which is calculated as a product of shear rate and material viscosity. Breaking up of an agglomerate will occur only when the shear stress exceeds the strength of the particle.

An important aspect of mixing studies on the torque rheometer is the temperature dependency of the mixing process. In general, the viscosity of a polymer decreases as temperature increases and vice versa. Properties of the material also change depending on temperature. Hence knowledge of the temperature dependency of viscosity is important. Large variations of viscosity for a certain range of temperature means that the material is thermally unstable (i.e., requires large activation energy). This kind of material has to be processed with accurate temperature control. Figure 10 shows the torque versus temperature curve obtained from the microprocessor-controlled torque rheometer to see the temperature dependency of viscosity-related torque on an EPDM material.

The MIM along with the torque rheometer can be used to simulate and optimize a variety of processing applications/problems. Some typical but by no means inclusive examples, are in studying fusion characteristics, examining stability and processability, color or thermal stability testing, examining the gelation of plastisols, in developing criteria for the selection of blowing agents for foam products, compound formulation optimization, studying the scorch and cure characteristics of rubber compounds, and in studying the cure characteristics of thermosets. An example is given below.

Lubricants play an important role in processing and in the properties of the final product. The lubricants also effect the fusion of the polymer materials.   That is, internal lubricants reduce melt viscosity, while external lubricants reduce friction between the melt and the hot melt parts of the processing equipment and prevent sticking, controlling the fusion of the resin. Figure 11 illustrates the results of an experiment aimed at studying the fusion characteristics of PVC. The level of external lubricant used in the formulation affects the fusing time between points L and F on the curve. The higher the level of external lubricant in the formulation, the longer the fusion time will be.

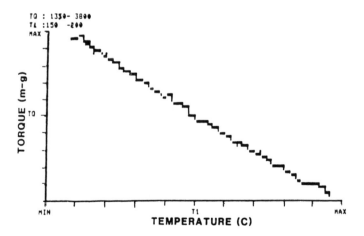

Figure 10. Cross plot of torque and temperature from rheocord.

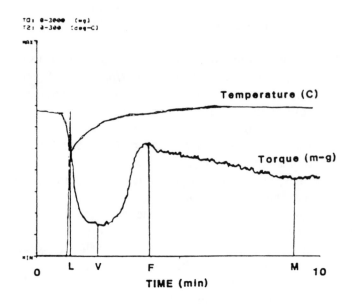

Figure 11.   Results of PVC fusion study.

If an unnecessarily high level of external lubricant is used in the formulation, it will take a longer period of time to melt the material in processing, which results in reducing production, increasing energy consumption, and poor products.   Meanwhile, if too low a level of external lubricant is used, the material will melt too early in the processing equipment, which may result in degradation in the final product. Therefore, selecting the optimum amount of external lubricant is a must for improvement of processing and for good-quality products.

One of the most frequently applied tests is in the study of additive incorporation and compounding.   All of the additives used in a formulation must be incorporated in the major component, and the components should be in a stable molecular arrangement.   Figure 12 illustrates a test result for incorporation of minor components to the major component as well as homogeneous compound after the additives are incorporated.   The test was performed with an EPDM rubber and reblended additives.   The EPDM was loaded into the mixer and mixed

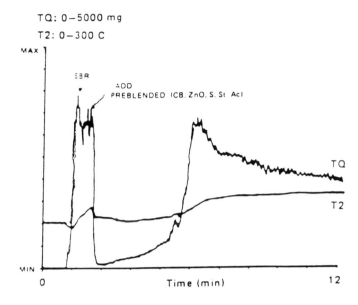

Figure 12. Development of incorporation time.

for 30 sec. Torque values immediately dropped sharply and increased as the additives incorporated.   When the ingredients were fully incorporated, a second torque peak was observed and finally stabilized when the material was homogeneously compounded. The second peak is called the "incorporation peak."   If hard fillers are added to the polymer, torque increases sharply and generates the second peak. This can be seen when carbon black is incorporated.   The time from the addition of the minor components resulting in the incorporation peak is referred to as the "incorporation time" and is critical to standard batch compounding operations with rubbers.

To this point, emphasis has been on applications testing where the torque rheometer has been retrofitted with a MIM. Another common practice is to incorporate a screw extruder. Solids conveying, melting, mixing, and pumping are the major functions of polymer processing extruders. The single-screw extruder is the most widely used machine to perform these functions. The plasticating extruder has three distinct regions: solids conveying zone, transition (melting) zone, and pumping zone.

The unit can be fed polymer in the particulate solids form or as strips, as in the case of rubber extrusion. The solids (usually in pellet or powder form) in the hopper flow by gravity into the screw channel, where they are conveyed through the solids conveying section. They are compressed by a drag-induced mechanism in the transition section. In other words, melting is accomplished by heat transfer from the heated barrel surface and by mechanical shear heating.

Simulation of the extrusion process in the laboratory is one of the most important applications of the torque rheometer in conjunction with single-screw extruders. Figure 13 illustrates simulations of widely used extrusion processes in the industries.

It is important for the process engineer to know the rheological properties of a material since the properties dominate the flow of the material in extrusion processes and also dominates the physical and mechanical properties of the extrudates. Therefore, it is also important to measure the properties utilizing a similar miniaturized extruder in the laboratory so that a process engineer knows the flow properties in the system by simulating the production line. Also, it is desirable to know the flow properties of a material to be processed in the range of shear rates of equipments to be used.

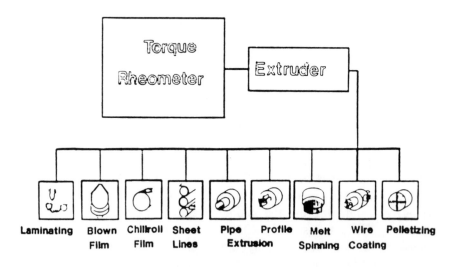

Figure 13. Application of torque rheometry to studying extrusion operations.

6   CHEMICAL ANALYSIS OF POLYMERS

Chemical analysis of polymer materials is difficult because of the large number and types of such materials and because of modification and compounding practices of conventional polymers.

For exact identification of polymers it is important for the samples to be in the form of pure products without incorporated additives such as plasticizers, fillers, or stabilizers. One must separate additives by extraction or reprecipitation before identification. The solvents or mixtures of solvent and precipitant are substance-specific and should be chosen separately for each case.

In order to determine the quantitative composition of a polymer, following sequence of operations is followed:

- Comminution of the polymer sample.
- Separation of additives.
- Qualitative and quantitative investigation of the additives.
- Identification and quantitative analysis of isolated polymer samples.

COMMINUTION, SEPARATION, AND IDENTIFICATION

Mechanical comminution of the sample is performed because the composition of polymers frequently shows inhomogeneities despite good processing. Furthermore, some tests (e.g., monomer content and H_2O content) depend on the sample.

Products are comminuted with cutting tools such as shears, knives, or razor blades. Drilling, milling, etc. are also suitable. A smaller particle diameter can also be obtained by grinding precooled samples in mills. Depending on the elasticity characteristics of the sample, either dry ice or liquid nitrogen can be effectively used.

97

After comminution, samples must be conditioned over phosphorus pentoxide at room temperature in a desiccator. In all mechanical loads and also in low temperature treatment, decomposition processes which may affect polymoleallarity and sample composition must be accounted for. Since the analytical result may be affected, attention must be paid to the reproducibility and uniformity of the comminution processes.

Plasticizers can be separated by extraction with diethyl ether. Stabilizers based on pure organic or organo-metallic compounds may only be partially separated. Extraction time depends on particle size and on the amount of plasticizer in the sample.

For the quantitative determination of the mass of plasticizer about 1 to 2 g of the comminuted sample are weighed and extracted with anhydrous diethyl ether in a Soxhlet apparatus. After distilling off the ether and drying the extract at 105°C to constant weight, the amount of ether soluble components is calculated from the difference in weight of the extraction flask before and after the extraction.   Preparative separation of the plasticizer before identification of the polymer is performed in an analogous manner.

Plasticizers include the esters of a few aliphatic and aromatic mono and dicarboxylic acids, aliphatic and aromatic phosphorus acid esters, ethers, alcohols, ketones, amines, amides, and non-polar and chlorinated hydrocarbons. These additives are used in various mixtures. For their separation and qualitative detection, thin-layer chromatography (TLC) is preferred. Usually Kieselgur plates, 0.25 mm thick, activated at 110°C for 30 min, in the saturated vapor are used. Methylene chloride and mixtures of diisopropyl ether/petether at temperatures between 40 to 60°C have been successfully used as the mobile phase. Refer to Table 1.

Under selected conditions, polymer plasticizers remain at the starting point. Although variation of the mobile and stationary phases and detection reactions allows selectivity of TLC to be increased, methods such as gas chromatography must be used for more complex plasticizer mixtures.

The gas-chromatographic separation of plasticizers can be effected directly or after conversion to low boiling point compounds. This is achieved by a transesterification reaction with methanol or diazomethane. After separation of the plasticizers mixtures with liquid chromatography, identification by spectroscopic methods is possible.

Inorganic fillers are, in general, insoluble in organic solvents. They can be quantitatively separated from the soluble polymers by

TABLE 1	
REAGENTS FOR THE DETECTION OF PLASTICIZERS IN TLC	
Type of Plasticizer	**Reagent**
Phthalates, adipates	Ethanolic resorcinol solution (20%) with 1% zinc chloride (10-min at 100°C); subsequently 4 N H_2SO_4 (20 min at 120°C)
Phosphates	Ammonium molybdate solution made from 3 g of ammonium molybdate in 20 m/ of perchloric acid (40%) and 5 m/ of concentrated hydrochloric acid in 200 m/ of water (10 min 100°C); subsequently saturated hydrazine sulphate solution (20 min 110°C)
Citrates	Ethanolic vanillin solution (20%) (10 min 80°C); subsequently 4 N H_2SO_4 (10 to 20 min 110°C)

centrifugation using solutions of 5 percent by weight in suitable solvents and subsequent pouring-off of the liquid.   Dissolution has to be accelerated by shaking or stirring.   The conditions of centrifugation depend on the density and particle size of the fillers.   Carbon black cannot be separated completely, even under the most severe centrifugation and special methods are available for its quantitative separation.   The same applies to the fillers in insoluble polymers, for example, in thermosets and unsaturated polyesters.   If required the fillers may be identified using the normal methods of qualitative inorganic analysis.

For the quantitative determination of fillers, a small amount of the polymer material is weighed and dissolved in solvent.   After centrifuging at the solution is pipetted off from the centrifugate.   The residue is suspended once more in the same solvent, centrifuges again and after isolation is washed once more with the solvent and then several times with methanol.   The residue is dried to constant weight and then the solid matter is determined gravimetrically. Table 2 reports the solubility of selected polymers.

STABILIZER IDENTIFICATION

Classification into heat stabilizers. Light stabilizers (also known as UV absorbers) and antioxidants is indicative of the great number of compounds which have gained importance. Among the most important heat stabilizers are basic salts of heavy metals, metal salts of organic acids, and nitrogenous organic compounds. Common antioxidants are phenols, aromatic amines, and benzimidazoles. UV absorbers are substances which absorb strongly in the short-wavelength range but are transparent at wave numbers < 25000 cm^{-1} so that the stabilized material does not show any coloration. The hydroxybenzophenone derivatives, salicyl esters, and benzotriazoles are examples.

Identification of stabilizers is complex because of their great number and the small amounts usually present. Added to this is the difficulty that stabilizers take part in transfer or rearrangement reactions during molding processes so that only a portion of the stabilizer is found unchanged in the finished product. Because of the intense toxicity of some of these decomposition products, their detection is of particular importance.

Stabilizers are identified after separation by solid-liquid extraction or after the removal of the polymer by precipitation from the diluted solution. Some extraction solvents for the most important stabilizers and polymers are given in Table 3.

TLC is the preferred separation method because of its high separation efficiency, rapidity, and large variety of detection possibilities. Usually 0.5 mm thick silica-gel-G-plates are used, activated at 120°C for 30 min. in a supersaturated atmosphere. Well-known of poly techniques such as multiple separation in opposite or parallel direction allow the selectivities to be further increased. The selection of an appropriate mobile phases determines the efficiency of separation. Advantage is taken of specific interactions and also of reactivity with the stabilizers under investigation.

TABLE 2

SOLUBILITY OF SELECTED POLYMERS

Polymer	Soluble	Insoluble
Polyvinyl chloride	Dimethyl formamide, tetrahydrofuran, cyclohexanone	Alcohols, hydrocarbons, butyl acetate
Polyvinylidene chloride	Ketones, tetrahydrofuran, dioxan, butyl acetate	Alcohols, hydrocarbons
Polyamides	Phenols, formic acid, trifluoroethanol	Alcohols, hydrocarbons, esters
Polyethylene	Tetralin, decalin, xylene, dichloroethylene at temperatures $> 100°C$	Alcohols, petrol, esters
Polystyrene	Ethyl acetate, benzene, acetone, chloroform, methylene dichloride	Alcohols, water
Polyvinyl acetate	Aromatic hydrocarbons, ketones, chlorinated hydrocarbons, alcohols	Petrol
Polyvinyl alcohol	Formamide, water	Ether, alcohols, petrol, benzene, esters, ketones, hydrocarbons
Polyurethanes	Dimethyl formamide	Ether, alcohols, petrol, benzene, water
Polyacrylonitrile	Dimethyl formamide, nitrophenols	Esters, alcohols, ketones, hydrocarbons
Polyester	Phenols, nitrated hydrocarbons, acetone, benzyl alcohol	Esters, alcohols, hydrocarbons
Aminoplastics	Benzylamine (160°C), ammonia	Organic solvents
Phenoplastics	Benzylamine (200°C)	
Cellulose, regenerated	SCHWEIZER's reagent	Organic solvents
Cellulose ester	Esters, ketones	Aliphatic hydrocarbons
Polybutadiene	Benzene	Alcohols, petrol, esters
Polyisoprene		Ketones
Polyisobutylene	Ether, petrol	Alcohols, esters
Polymethacrylic acid	Aromatic hydrocarbons, esters, ketones, chlorinated hydrocarbons	Ether, alcohols, aliphatic hydrocarbons

<table>
<tr><td colspan="3" align="center">TABLE 3

STABILIZER IDENTIFICATION</td></tr>
<tr><td align="center">Polymer</td><td align="center">Stabilizers to be Extracted</td><td align="center">Extracting Agent</td></tr>
<tr><td>Polyvinyl chloride</td><td>Organotin stabilizers</td><td>Heptane: glacial Acetic acid 1:1</td></tr>
<tr><td>Polyvinyl chloride</td><td>N-containing organic stabilizers</td><td>Methanol or diethyl ether</td></tr>
<tr><td>Polyethylene</td><td>Antioxidants</td><td>Chloroform</td></tr>
<tr><td>Polyoxymethylene</td><td>Phenolic antioxidants</td><td>Chloroform</td></tr>
<tr><td>Rubbers</td><td>Stabilizers; accelerators</td><td>Boiling acetone; boiling water</td></tr>
</table>

POLYMER IDENTIFICATION

Polymer identification starts with a series of preliminary tests. In contrast to low molecular weight organic compounds, which are frequently satisfactorily identified simply by their melting or boiling point, molecular weight and elementary composition, precise identification of polymers is difficult by the presence of copolymers, the statistical character of the composition, macromolecular properties and, by potential polymeric-analogous reactions. Exact classification of polymers is not usually possible from a few preliminary tests. Further physical data must be measured and specific reactions must be carried out in order to make a reliable classification. The efficiency of physical methods such as IR spectroscopy and NMR spectroscopy as well as pyrolysis gas chromatography makes them particularly important.

One method of analysis is pyrolysis, which is the application of thermal energy, causing covalent bonds to be broken. Fragments are produced whose chain length and structure are dependent upon the temperature, on the one hand, and on the type of bonds, on the other. If oxygen is present, oxidation occurs at the same time which may lead to ignition. Since pyrolysis itself is an endothermic reaction, the required energy must be supplied, by an external heat source. The subsequent oxidation process is exothermic. If sufficient energy is released, the sample will burn spontaneously. If the temperature drops after removal of the flame, the polymer will be self-extinguishing.

Behavior in an open flame can easily be observed by holding about 0.1 . . . 0.2g of sample with a suitable implement in the outer edge of a small Bunsen flame.

To test behavior during dry heating, about 0.1g of the sample is carefully heated in a 60 mm long glow tube with a diameter of 6 mm over a small flame. If heating is too vigorous, the characteristic phenomena can no longer be observed.

Depolymerization is a special case of thermal degradation. It can be observed particularly in polymers based on a, a'-disubstituted monomers. In these, degradation is a reversal of the synthesis process. It is a chain reaction during which the monomers are regenerated by an unzipping mechanism. This is due to the low polymerization enthalpy of these polymers. For the thermal fission of polymers with secondary and tertiary C-atoms, higher energies are required. In these cases elimination reactions occur. This can be seen very clearly in PVC and PVAC.

Depolymerization is in functional correlation with the molecular weight distribution and with the type of terminal groups which are formed in chain initiation and chain termination.

Depolymerization, elimination, and statistical chain-scission reactions can be used for polymer analysis. When the monomer is the main degradation product obtained, it can easily be identified by boiling point and refractive index.

Elimination and chain-scission reactions provide characteristics pyrograms which can often be identified by gas chromatography or IR spectroscopy.

For testing depolymerization behavior, about 0.2 to 0.3g of the polymeric substance is carefully and gently heated to a maximum of 500°C in a small distillation flask. The distillate, is collected in a receiver and its boiling point and refractive index are determined.

APPENDIX A

ABBREVIATIONS OF POLYMERS

ABS	Acrylonitrile-butadiene-styrene
AN	Acrylonitrile
CA	Cellulose acetate
CAB	Cellulose acetate butyrate
CAP	Cellulose acetate propionate
CN	Cellulose nitrate
CP	Cellulose propionate
CPE	Chlorinated polyethylene
CPVA	Chlorinated polyvinyl chloride
CTFE	Chlorotrifluoroethylene
DAP	Diallyl phthalate
EC	Ethyl cellulose
ECTFE	Poly(ethylene-chlorotrifuloroethylene)
EP	Epoxy
EPDM	Ethylene-propylene-diene monomer
EPR	Ethylene propylene rubber
EPS	Expanded polystyrene
ETFE	Ethylene/tetrafluoroethylene copolymer
EVA	Ethylene-vinyl acetate
FEP	Perfluoro (ethylene-propylene) copolymer
FRP	Fiberglass-reinforced polyester
HDPE	High-density polyethylene
HIPS	High-impact polystyrene
HMWPE	High-molecular-weight polyethylene
LDPE	Low-density polyethylene
MF	Melamine-formaldehyde
PA	Polyamide
PAPI	Polymethylene polyphenyl isocyanate
PB	Polybutylene
PBT	Polybutylene terephthalate (thermoplastic polyester)

PC	Polycarbonate
PE	Polyethylene
PES	Polyether sulfone
PET	Polyethylene terephthalate
PF	Phenol-formaldehyde
PFA	Polyfluoro alkoxy
PI	Polyimide
PMMA	Polymethyl methacrylate
PP	Polypropylene
PPO	Polyphenylene oxide
PS	Polystyrene
PSO	Polysulfone
PTFE	Polytetrafluoroethylene
PTMT	Polytetramethylene terephthalate (thermoplastic polyester)
PU	Polyurethane
PVA	Polyvinyl alcohol
PVAC	Polyvinyl acetate
PVC	Polyvinyl chloride
PVDC	Polyvinylidene chloride
PVDF	Polyvinylidene fluoride
PVF	Polyvinyl fluoride
TFE	Polytelrafluoroethylene
SAN	Styrene-acrylonitrile
SI	Silicone
TP	Thermoplastic Elastomers
TPX	Polymethylpentene
UF	Urea formaldehyde
UHMWPE	Ultrahigh-molecular-weight polyethylene
UPVC	Unplasticized polyvinyl chloride

APPENDIX B

GLOSSARY OF POLYMERS AND TESTING

abrasion resistance--ability of material to withstand mechanical action such as rubbing, scraping, or erosion that tends to progressively remove material from its surface.

accelerated aging--test in which conditions are intensified in order to reduce the time required to obtain a deteriorating effect similar to one resulting from normal service conditions.

accelerated weathering--test in which the normal weathering conditions are accelerated by means of a device.

aging--process of exposing plastics to natural or artificial environmental conditions for prolonged period of time.

amorphous polymers--polymeric materials that have no definite order or crystallinity. Polymer molecules are arranged in completely random fashion.

apparent density--(bulk density) weight of unit volume of material including voids (air) inherent in the material.

arc resistance--ability of plastic to resist the action of a high voltage electrical arc, usually in terms of time required to render the material electrically conductive.

birefringence (double refraction)--the difference between index of refraction of light in two directions of vibration.

brittle failure--failure resulting from inability of material to absorb energy, resulting in instant fracture upon mechanical loading.

brittleness temperature--temperature at which plastics and elastomers exhibit brittle failure under impact conditions.

bulk factor--ratio of volume of any given quantity of the loose plastic material to the volume of the same quantity of the material after molding or forming. It is a measure of volume change that may be expected in fabrication.

burst strength--the internal pressure required to break a pressure vessel such as a pipe or fitting. The pressure (and therefore the burst strength) varies with the rate of pressure buildup and the time during which the pressure is held.

capillary rheometer--instrument for measuring the flow properties of polymer melts. Comprised of a capillary tube of specified diameter and length, means for applying desired pressures to force molten polymer through the capillary, means for maintaining the desired temperature of the apparatus, and means for measuring differential pressures and flow rates.

chalking--a whitish, powdery residue on the surface of a material caused by material degradation (usually from weather).

Charpy impact test--a destructive test of impact resistance, consisting of placing the specimen in a horizontal position between two supports then striking the specimen with a pendulum striker swung from a fixed height. The magnitude of the blow is increased until specimen breaks. The result is expressed in in-lb or ft-lb of energy.

CIE (Commission Internationale De L Eclairage)--international commission on illuminants responsible for establishing standard illuminants.

coefficient of thermal expansion--fractional change in length or volume of a material for unit change in temperature.

colorimeter--instrument for matching colors with results approximately the same as those of visual inspection, but more consistently.

compressive strength--maximum load sustained by a test specimen in a compressive test divided by original cross section area of the specimen.

conditioning--subjecting a material to standard environmental and/or stress history prior to testing.

continuous use temperature--maximum temperature at which material may be subjected to continuous use without fear of premature thermal degradation.

crazing--undesirable defect in plastic articles, characterized by distinct surface cracks or minute frostlike internal cracks, resulting from stresses within the article. Such stresses result from molding shrinkage, machining, flexing, impact shocks, temperature changes or action of solvents.

creep--due to viscoelastic nature, a plastic subjected to a load for a period of time tends to deform more than it would from the same load released immediately after application. The degree of this deformation is dependent on the load duration. Creep is the permanent deformation resulting from prolonged application of stress below the elastic limit. Creep at room temperature is called cold flow.

creep rupture strength--stress required to cause fracture in a creep test.

crosslinking--the setting up of chemical links between the molecular chains. When extensive, as in most thermosetting resins, crosslinking makes one infusible super-molecule of all the chains. Crosslinking can be achieved by irradiation with high energy electron beams or by chemical crosslinking agents.

crystallinity--state of molecular structure attributed to existence of solid crystals with a definite geometric form. Such structures are characterized by uniformity and compactness.

cup flow test--test for measuring the flow properties of thermosetting materials. A standard mold is charged with preweighed material, and the mold is closed using sufficient pressure to form a required cup. Minimum pressures required to mold a standard cup and the time required to close the mold fully are determined.

cup viscosity test--test for making flow comparisons under strictly comparable conditions. The cup viscosity test employs a cup-shaped gravity device that permits the timed flow of a known volume of liquid passing through an orifice located at the bottom of the cup.

density--weight per unit volume of a material expressed in grams per cubic centimeter, pounds per cubic foot, etc.

dielectric constant (permititivity)--ratio of the capacitance of a given configuration of electrodes with a material as dielectric to the capacitance of the same electrode configuration with a vacuum (or air for most practical purposes) as the dielectric.

differential scanning calorimetry (DSC)--thermal analysis technique that measures the quantity of energy absorbed or evolved (given of by a specimen in calories as its temperature is changed.

dimensional stability--ability to retain the precise shape in which it was molded, fabricated or cast.

dissipation factor--ratio of the conductance of a capacitor in which the material is dielectric to its susceptance, or the ratio of its parallel reactance to its parallel resistance. Most plastics have a low dissipation factor, a desirable property because it minimizes the waste of electrical energy as heat.

drop impact test--impact resistance test in which a predetermined weight is allowed to fall freely onto the specimen from varying heights. The energy absorbed by the specimen is measured and expressed in in-lb or ft-lb.

ductility--extent to which a material can sustain plastic deformation without fracturing.

durometer hardness--measure of the indentation hardness of plastics. It is the extent to which a spring-loaded steel indentor protrudes beyond the pressure foot into the material.

elongation--the increase in length of a test specimen produced by a tensile load.   Higher elongation indicates higher ductility.

embrittlement--reduction in ductility due to physical or chemical changes.

environmental stress cracking--the susceptibility of a thermoplastic article to crack or craze formation under the influence of certain chemicals and stress.

extensometer--instrument for measuring changes in linear dimensions (also called strain gauge).

extrusion plastometer (rheometer)--a type of viscometer used for determining the melt index of a polymer. Comprised of a vertical cylinder with two longitudinal bored holes (one for measuring temperature and one for containing the specimen, the latter having an orifice of stipulated diameter at the bottom and a plungering from the top).   The cylinder is heated by external bands and weight is placed on the plunger to force the polymer specimen through the orifice. The result is reported in grams/10 min.

fadometer--an apparatus for determining the resistance of materials to fading by exposing them to ultraviolet rays of approximately the same wavelength as those found in sunlight.

fatigue failure--the failure or rupture of a plastic under repeated cyclic stress, at a point below the normal static breaking strength.

fatigue limit--the stress below which a material can be stressed cyclically for an infinite number of times without failure.

fatigue strength--the maximum cyclic stress a material can withstand for a given number of cycles before failure.

flammability--measure of the extent to which a material will support combustion.

flexural modulus--ratio of the applied stress on a test specimen in flexure to the corresponding strain in the outermost fiber of the specimen. Flexural modulus is the measure of relative stiffness.

flexural strength--the maximum stress in the outer fiber at the moment of crack or break.

foamed plastics (cellular plastics)--plastics with numerous cells disposed throughout its mass. Cells are formed by a blowing agent or by the reaction of the constituents.

gel permeation chromatography (GPC)--column chromatography technique employing a series of columns containing closely packed rigid gel particles. The polymer to be analyzed is introduced at the top of the column and then is eluted with a solvent. The polymer molecules diffuse through the gel at rates depending on their molecular size. As they emerge from the columns, they are detected by differential refractometer coupled to a chart recorder, on which a molecular weight distribution curve is plotted.

gel point--the stage at which liquid begins to gel, that is, exhibits pseudoelastic properties.

hardness--the resistance of plastic materials to compression and indentation. Brinnel hardness and shore hardness are major methods of testing this property.

haze--the cloudy or turbid aspect of appearance of an otherwise transparent specimen caused by light scattered from within the specimen or from its surface.

Hooke's Law--stress is directly proportional to strain.

hoop stress--the circumferential stress in a material of cylindrical form subjected to internal or external pressure.

hygroscopic--material having the tendency to absorb moisture from air. Plastics, such as nylons and ABS, are hygroscopic and must be dried prior to molding.

hysteresis--the cyclic noncoincidence of the elastic loading and the unloading curves under cyclic stressing. The area of the resulting elliptical hysteresis loop is equal to the heat generated in the system.

impact strength--energy required to fracture a specimen subjected to shock.

impact test--method of determining the behavior of material subjected to shock loading in bending or tension. The quantity usually measured is the energy absorbed in fracturing the specimen in a single blow.

indentation hardness--resistance of a material to surface penetration by an indentor. The hardness of a material as determined by the size of an indentation made by an indenting tool under a fixed load, or the load necessary to produce penetration of the indentor to a predetermined depth.

index of refraction--ratio of velocity of light in vacuum (or air) to its velocity in a transparent medium.

infrared analysis--technique used for polymer identification. An infrared spectrometer directs infrared radiation through a film or layer of specimen and measures the relative amount of energy absorbed by the specimen as a function of wavelength or frequency of infrared radiation. The chart produced is compared with correlation charts for known substances to identify the specimen.

inherent viscosity--in dilute solution viscosity measurements, inherent viscosity is the ratio of the natural logarithm of the relative viscosity to the concentration of the polymer in grams per 100 ml of solvent.

ISO--abbreviation for the International Standards Organization.

isochronous (equal time) stress-strain curve--a stress-strain curve obtained by plotting the stress vs.

corresponding strain at a specific time of loading pertinent to a particular application.

IZOD impact test--method for determining the behavior of materials subjected to shock loading. Specimen supported as a cantilever beam is struck by a weight at the end of a pendulum. Impact strength is determined from the amount of energy required to fracture the specimen. The specimen may be notched or unnotched.

melt index test--measures the rate of extrusion of a thermoplastic material through an orifice of specific length and diameter under prescribed conditions of temperature and pressure. Value is reported in grams per 10 minutes for specific condition.

modulus of elasticity (elastic modulus, Young's modulus)--the ratio of stress to corresponding strain below the elastic limit of a material.

molecular weight--the sum of the atomic weights of all atoms in a molecule. In high polymers, the molecular weight of individual molecules varies widely, therefore, they are expressed as weight average or number average molecular weight.

molecular weight distribution--the relative amount of polymers of different molecular weights that comprise a given specimen of a polymer.

monomer--(monomer single-unit) a relatively simple compound that can react to form a polymer (multiunit) by combination with itself or with other similar molecules or compounds.

necking--the localized reduction in cross section that may occur in a material under stress.   Necking usually occurs in a test bar during a tensile test.

notch sensitivity--measure of reduction in load-carrying ability caused by stress concentration in a specimen. Brittle plastics are more notch sensitive than ductile plastics.

orientation--the alignment of the crystalline structure in polymeric materials so as to produce a highly uniform structure.

oxygen index--the minimum concentration of oxygen expressed as a volume percent, in a mixture of oxygen and nitrogen that will just support flaming combustion of a material initially at room temperature under the specified conditions.

peak exothermic temperature-- the maximum temperature reached by reacting thermosetting plastic composition is called peak exothermic temperature.

photoelasticity--experimental technique for the measurement of stresses and strains in material objects by means of the phenomenon of mechanical birefringence.

Poisson's ratio--ratio of lateral strain to axial strain in an axial loaded specimen. It is a constant that relates the modulus of rigidity to Young's modulus.

polarized light--polarized electromagnetic radiation whose frequency is in the optical region.

polarizer--a medium or a device used to polarize the incoherent light.

polymerization--a chemical reaction in which the molecules of monomers are linked together to form polymers.

proportional limit--the greatest stress that a material is capable of sustaining without deviation from porportionality of stress and strain (Hooke's Law).

relative viscosity--ratio of kinematic viscosity of a specified solution of the polymer to the kinematic viscosity of the pure solvent.

rheology--the science dealing with the study of material flow.

Rockwell hardness--index of indentation hardness measured by a steel ball indentor.

secant modulus--the ratio of total stress to corresponding strain at any specific point on the stress-strain curve.

shear rate--the overall velocity over the cross section of a channel with which molten or fluid layers are gliding along each other or along the wall in laminar flow.

shear strength--the maximum load required to shear a specimen in such a volume manner that the resulting pieces are completely clear of each other.

SPE--abbreviation for Society of Plastics Engineers.

specific gravity--the ratio of the weight of the given volume of a material to that of an equal volume of water at a stated temperature.

spectrophotometer--an instrument that measures transmission or apparent reflectance of visible light as a function of wavelength, permitting accurate analysis of color or accurate comparison of luminous intensities of two sources of specific wavelengths.

specular gloss--the relative luminous reflectance factor of a specimen at the specular direction.

SPI--abbreviation for Society of Plastics Industry.

spiral flow test--a method for determining the flow properties of a plastic material based on the distance it will flow under controlled conditions of pressure and temperature along the path of a spiral cavity using a controlled charge mass.

strain--the change in length per unit of original length, usually expressed in percent.

stress--the ratio of applied load of the original cross sectional area expressed in pounds per square inch.

stress concentration--the magnification of the level of applied stress in the region of a notch, crack, void, inclusion or other stress risers.

stress optical sensitivity--the ability of materials to exhibit double refraction of light when placed under stress is referred to a stress-optical sensitivity.

stress relaxation--the gradual decrease in stress with time under a constant deformation (strain).

stress-strain diagram--graph of stress as a function of strain. It is constructed from the data obtained in any mechanical test where a load is applied to a material and continuous measurements of stress and strain are made simultaneously.

tensile impact energy--the energy required to break a plastic specimen in tension by a single swing of a calibrated pendulum.

tensile strength--ultimate strength of a material subjected to tensile loading.

thermal conductivity--the ability of a material to conduct heat. The coefficient of thermal conductivity is expressed as the quantity of heat that passes through a unit cube of the substance in a given unit of time when the difference in temperature of the two faces is 1 degree.

thermogravimetric analysis (TGA)--a testing procedure in which changes in the weight of a specimen are recorded as the specimen is progressively heated.

thermomechanical analysis (TMA)--a thermal analysis technique consisting of measuring physical expansion or contraction of a material or changes in its modulus or viscosity as a function of temperature.

thermoplastic--a class of plastic material that is capable of being repeatedly softened by heating and hardened by cooling. ABS, PVC, polystyrene, polyethylene, etc. are thermoplastic materials.

thermosetting plastics--a class of plastic materials that will undergo a chemical reaction by the action of heat, pressure, catalyst, etc. leading to a relatively infusible, nonreversible state. Phenolics, epoxies, and alkyds are examples of typical thermosetting plastics.

torsion--stress caused by twisting a material.

torsion pendulum--an equipment used for determining dynamic mechanical plastics.

toughness--the extent to which a material absorbs energy without fracture. The area under a stress-strain diagram is also a measure of toughness of a material.

tristimulus colorimeter--the instrument for color measurement based on spectral tristimulus values. Such an instrument measures color in terms of three primary colors: red, green, and blue.

ultrasonic testing--a nondestructive testing technique for detecting flaws in material and measuring thickness based on the use of ultrasonic frequencies.

ultraviolet--the region of the electromagnetic spectrum between the violet end of visible light and the x-ray region, including wavelengths from 100 to 3900 A. Photon of radiations in the UV area have sufficient energy to initiate some chemical reactions and to degrade some plastics.

vicat softening point--the temperature at which a flat-ended needle of one square mm circular or square cross section will penetrate a thermoplastic specimen to a depth of 1 mm under a specified load using a uniform rate of temperature rise.

viscometer--an instrument used for measuring the viscosity and flow properties of fluids.

viscosity--a measure of resistance of flow due to internal friction when one layer of fluid is caused to move in relationship to another layer.

water absorption--the amount of water absorbed by a polymer when immersed in water for stipulated period of time.

weathering--a term encompassing exposure of polymers to solar or ultraviolet light, temperature, oxygen, humidity, snow, wind, pollution, etc.

weatherometer--an instrument used for studying the effect of weather on plastics in accelerated manner using artificial light sources and simulated weather conditions.

yellowness index--measure of the tendency of plastics to turn yellow upon long term exposure to light.

yield point--stress at which strain increases without accompanying increase in stress.

yield strength--the stress at which a material exhibits a specified limiting deviation from the proportionality of stress to strain. Unless otherwise specified, this stress will be the stress at the yield point.

Young's modulus--the ratio of tensile stress to tensile strain below the proportional limit.

APPENDIX C

PROFESSIONAL AND TESTING ORGANIZATIONS

AMERICAN NATIONAL STANDARDS INSTITUTE (ANSI)

ANSI is a federation of standards compiled from commerce and industry, professional, trade, consumer, and labor organizations and government. ANSI helps to perform the following:

- Identifies the needs for standards and sets priorities for their completion.
- Assigns development work to competent and willing organizations.
- Sees to it that public interests, including those of the consumer, are protected and represented.
- Supplies standards writing organizations with effective procedures and management services to ensure efficient use of their manpower and financial resources and timely development of standards.
- Follows up to assure that needed standards are developed on time.

Another role is to approve standards as American National Standards when they meet consensus requirements. It approves a standard only when it has verified evidence presented by a standards developer that those affected by the standard have reached substantial agreement on its provisions. ANSI's other major roles are to represent U.S. interests in nongovernmental international standards work, to make national and international standards available, and to inform the public.

AMERICAN SOCIETY FOR TESTING AND MATERIALS (ASTM)

ASTM is a scientific and technical organization formed for "the development of standards on characteristics and performance of materials, products, systems and services, and the promotion of related knowledge." ASTM is the world's largest source of voluntary consensus standards. The society operates through more than 135 main technical committees with 1550 subcommittees. These committees function in prescribed fields under regulations that ensure balanced representation among producers, users, and general interest participants. The society currently has 28,000 active members, of whom approximately 17,000 serve as technical experts on committees, representing 76,200 units of participation.

Membership in the society is open to all concerned with the fields in which ASTM is active. An ASTM standard represents a common viewpoint of those parties concerned with its provisions namely, producers, users, and general interest groups. It is intended to aid industry, government agencies, and the general public. The use of an ASTM standard is voluntary. It is recognized that, for certain work, ASTM specifications may be either more or less restrictive than needed. The existence of an ASTM standard does not preclude anyone from manufacturing, marketing, or purchasing products, or using products, processes, or procedures not conforming to the standard. Because, ASTM standards are subject to periodic reviews and revision, it is recommended that all serious users obtain the latest revision. A new edition of the Book of Standards is issued annually. On the average about 30 percent of each part is new or revised.

FOOD AND DRUG ADMINISTRATION (FDA)

The Food and Drug Administration is an U.S. government agency of the Department of Health and Human Services. The FDA's activities are directed toward protecting the health of the nation against impure and unsafe foods, drugs, cosmetics, and other potential hazards.

The plastics industry is mainly concerned with the Bureau of Foods which conducts research and develops standards on the composition, quality, nutrition, and safety of foods, food additives, colors and cosmetics, and conducts research designed to improve the detection, prevention, and control of contamination. The FDA is concerned about indirect additives. Indirect additives are those substances capable of migrating into food from contacting plastic materials. Extensive tests are carried out by the FDA before issuing safety clearance to any plastic material that is to be used in food contact applications. Plastics used in medical devices are tested with extreme caution by the FDA's Bureau of Medical Devices which develops FDA policy regarding safety and effectiveness of medical devices.

NATIONAL BUREAU OF STANDARDS (NBS)

National Bureau of Standards overall goal is to strengthen and advance the nation's science and technology and to facilitate their effective application for public benefit.

The bureau conducts research and provides a basis for the nation's physical measurement system, scientific and technological services for industry and government, a technical basis for increasing productivity and innovation, promoting international competitiveness in American industry, maintaining equity in trade, and technical services, promoting public safety. The bureau's technical work is performed by the National Measurement Laboratory, the National Engineering Laboratory and the Institute for Compute Sciences and Technology.

NATIONAL ELECTRICAL MANUFACTURERS ASSOCIATION (NEMA)

The National Electrical Manufacturers Association consists of manufacturers of equipment and apparatus for the generation, transmission, distribution, and utilization of electric power. The membership is limited to corporations, firms, and individuals actively engaged in the manufacture of products included within the product scope of NEMA product subdivisions.

NEMA develops product standards covering such matters as nomenclature, ratings, performance, testing, and dimensions. NEMA is also actively involved in developing National Electrical Safety Codes and advocating their acceptance by state and local authorities. Along with a monthly news bulletin, NEMA also publishes manuals, guidebooks, and other material on wiring, installation of equipment, lighting, and standards. The majority of NEMA standardization activity is in cooperation with other national organizations. The manufacturers of wires and cables, insulating materials, conduits, ducts, and fittings are required to adhere to NEMA standards by state and local authorities.

NATIONAL FIRE PROTECTION ASSOCIATION (NFPA)

National Fire Protection Association has the objective of developing, publishing, and disseminating standards intended to minimize the possibility and effect of fire and explosion.   NFPA's membership consists of individuals from business and industry, fire service, health care, insurance, educational, and government institutions.   NFPA conducts fire safety education programs for the general public and provides information on fire protection and prevention. Also provided by the association is the field service by specialists on flammable liquids, electricity, gases, and marine problems.

Each year, statistics on causes and occupancies of fires and deaths resulting from fire are compiled and published.   NFPA sponsors seminars on the Life Safety Codes, National Electrical Code, industrial fire protection, hazardous materials, transportation emergencies, and other related topics. NFPA also conducts research programs on delivery systems for public fire protection, arson, residential fire sprinkler systems, and other subjects. NFPA publications include National Fire Codes Annual, Fire Protection Handbook, Fire Journal and Fire Technology.

NATIONAL SANITATION FOUNDATION (NSF)

The National Sanitation Foundation, is an independent, nonprofit environmental organization of scientists, engineers, technicians, educators, and analysts.   NSF frequently serves as a trusted neutral agency for government, industry, and consumers, helping them to resolve differences and unite in achieving solutions to problems of the environment.

At NSF, a great deal of work is done on the development and implementation of NSF standards and criteria for health-related equipment. The majority of NSF standards relate to water treatment and purification equipment, products for swimming pool applications, plastic pipe for potable water as well as drain, waste, and vent (DWV) uses, plumbing components for mobil homes and recreational vehicles, laboratory furniture, hospital cabinets, polyethylene refuse bags and containers, aerobic waste treatment plants, and other products related to environmental quality.

Manufacturers of equipment, materials, and products that conform to NSF standards are included in official listings, and these producers are authorized to place the NSF seal on their products. Representatives from NSF regularly visit the plants of manufacturers to make certain that products bearing the NSF seal, do fulfill applicable NSF standards.

PLASTICS TECHNICAL EVALUATION CENTER (PLASTEC)

PLASTEC is one of 20 information analysis centers sponsored by the Department of Defense to provide the defense community with a variety of technical information services applicable to plastics, adhesives, and organic matrix composites. For the last 21 years, PLASTEC has served the defense community with authoritative information and advice in such forms as engineering assistance, responses to technical inquiries, special investigations, field trouble shooting, failure analysis, literature searches, state-of-the-art reports, data compilations, and handbooks. PLASTEC has also been heavily involved in standardization activities. In recent years, PLASTEC has been permitted to serve private industry.

The significant difference between a library and technical evaluation center is the quality of the information provided to the user. PLASTEC uses its database library as a means to an end to provide succinct and timely information which has been carefully evaluated and analyzed. Examples of the activity include recommendation of materials, counseling on designs, and performing trade-off studies between various materials, performance requirements, and costs. Applications are examined consistent with current manufacturing capabilities, and the market availability of new and old materials alike is considered. PLASTEC specialists can reduce raw data to the user's specifications and supplement them with unpublished information that updates and refines published data. PLASTEC works to spin-off the results of government-sponsored R&D to industry and similarly to utilize commercial advancements to the government's goal of highly sought technology transfer. PLASTEC has a highly specialized library to serve the varied needs of their own staff and customers.

PLASTEC offers a great deal of information and assistance to the design engineer in the area of specifications and standards on plastics. PLASTEC has complete visual search microfilm file and can display and print the latest issues of specifications, test methods, and standards from Great Britain, Germany, Japan, U.S.A., and International Standards Organization.   Military and federal specifications and standards and industry standards such as ASTM, NEMA, and UL are on file and can be quickly retrieved.

SOCIETY OF PLASTICS ENGINEERS (SPE)

The Society of Plastics Engineers promotes scientific and engineering knowledge relating to plastics. SPE is a professional society of plastics scientists, engineers, educators, students, and others interested in the design, development, production, and utilization of plastics materials, products, and equipment. SPE currently has over 22,000 members scattered among its 80 sections. The individual sections as well s the SPE main body arranges and conducts monthly meetings, conferences, educational seminars, and plant tours throughout the year. SPE also publishes Plastics Engineering, Polymer Engineering and Science, Plastics Composites, and the Journal of Vinyl Technology. The society presents a number of awards each year encompassing all levels of the organization, section, division, committee, and international.   SPE divisions of interest are color and appearance, injection molding, extrusion, electrical and electronics, thermoforming, engineering properties and structure, vinyl plastics, blow molding, medical plastics, plastics in building, decorating, mold making, and mold design.

SOCIETY OF PLASTICS INDUSTRY (SPI)

The Society of Plastics Industry is a major society, whose membership consists of manufacturers and processors of plastics materials and equipment. The society has four major operating units consisting of the Eastern Section, the Midwest Section, the New England Section and the Western Section.   SPI's Public Affairs Committee concentrates on coordinating and managing the response of the plastics industry to issues like toxicology, combustibility, solid waste, and energy. The Plastic Pipe Institute is one of the most active divisions, promoting the proper use of plastic pipes by establishing standards, test procedures, and specifications. Epoxy Resin Formulators Division has published over 30 test procedures and technical specifications.   Risk management, safety standards, productivity, and quality are a few of the major programs undertaken by the machinery division. SPI's other divisions include Expanded Polystyrene Division, Fluoropolymers Division, Furniture Division, International Division, Plastic Bottle Institute, Machinery Division, Molders Division, Mold Makers Division, Plastic Beverage Container Division, Plastic Packaging Plastic/Composites Institute, Structural Foam Division, Vinyl Siding Institute, Vinyl Formulators Division.

The National Plastics Exposition and Conference, held every three years by the Society of Plastic Industry, is one of the largest plastic shows in the world.

UNDERWRITERS LABORATORIES (UL)

Underwriters Laboratories, is a not-for-profit organization whose goods are to establish, maintain, and operate laboratories for the investigation of materials, devices, products equipment, constructions, methods, and systems with respect to hazards affecting life and property.

There are five testing facilities in the U.S. and over 200 inspection centers. More than 700 engineers and 500 inspectors conduct tests and follow-up investigations to insure that potential hazards are evaluated and proper safeguards provided. UL has six basic services it offers to manufacturers, inspection authorities, or government officials. These are product listing service, classification, service, component recognition service, certificate service, inspection service, and fact finding and research.

UL's Electrical Department is in charge of evaluating individual plastics and other products using plastics as components. Electrical Department evaluates consumer products such as TV sets, power tools, appliances, and industrial and commercial electrical equipment and components. In order for a plastic material to be recognized by UL it must pass a variety of UL tests including the UL 94 flammability test and the UL 746 series, short- and long-term property evaluation tests. When a plastic material is granted Recognized Component Status, a yellow card is issued. The card contains precise identification of the material including supplier, product designation, color, and its UL 94 flammability classification at one or more thicknesses. Also included are many of the property values such as temperature index, hot wire ignition, high-current arc ignition and arc resistance. These data also appear in the recognized component directory.

UL publishes the names of the companies who have demonstrated the ability to provide a product conforming to the established requirements, upon successful completion of the investigation and after agreement of the terms and condition of the listing and follow-up service. Listing signifies that production samples of the product have been found to comply with the requirements and that the manufacturer is authorized to use the UL's listing mark on the listed products which comply with the requirements.

UL's consumer advisory council was formed to advise UL in establishing levels of safety for consumer products to provide UL with additional user field experience and failure information in the field of product safety, and to aid in educating the general public in the limitations and safe use of specific consumer products.

GLOSSARY OF ENGINEERING AND MATERIALS TERMS

This appendix provides definitions and a source of basic information about polymers and chemical products - their properties, the processes by which they are made, the test methods used to assess product characteristics and assure product quality, and the types of equipment and materials that rely on feedstocks for their operation or manufacture. In most of the definitions, certain words and phrases are in *italics*; others are in **boldface**. *Italics* indicate that the term is defined separately under its own heading; **boldface** indicates that the term is of key importance in the definition or that it is a commonly used alternative word or phrase being described.

In many definitions, reference is made to ASTM methods. These are carefully controlled test methods developed by the American Society for Testing and Materials (ASTM), a scientific and technical organization founded in 1898 to establish standards for products, systems, and services.

absolute humidity--see *humidity*.

absolute pressure--see *pressure*.

absolute scale--see *temperature scales*.

absolute viscosity--the ratio of *shear stress* to *shear rate*. It is a fluid's internal resistance to flow. The common unit of absolute viscosity is the poise (see *viscosity*). Absolute viscosity divided by the fluid's density equals *kinematic viscosity*.

absorber oil--oil used to selectively absorb heavier hydrocarbon components from a gas mixture. Also called **wash oil** or **scrubber oil**.

absorption--the assimilation of one material into another: in petroleum refining, the use of an absorptive liquid to selectively remove components from a *process stream*.

AC--see *asphalt cement*.

acetylene--highly flammable hydrocarbon gas (C_2H_2) used in welding and cutting, and in plastics manufacture. Also a term for a series of unsaturated *aliphatic hydrocarbons*, each containing at least one triple carbon bond, the simplest member of the series being acetylene. The triple carbon bond makes acetylenes highly reactive. See *hydrocarbon, unsaturated hydrocarbon*.

$$H - C \equiv C - H$$
acetylene

acid--hydrogen-containing compound that reacts with metals to form salts, and with metallic oxides and bases to form a salt and water. The strength of an acid depends on the extent to which its molecules ionize, or dissociate, in water, and on the resulting concentration of hydrogen ions (H^-) in solution. Petroleum hydrocarbons, in the presence of oxygen and heat, can oxidize to form weak acids, which attack metals. See *corrosion*.

acidizing--treatment of underground oil-bearing formations with acid in order to increase production. Hydrochloric or other acid is injected into the formation and held there under pressure until it etches the rock, thereby enlarging the pore spaces and passages through which the oil flows. The acid is then pumped out and the well is swabbed and put back into production.

acid number--refining process for improving the color, odor, and other properties of white oils or lube stocks, whereby the unfinished product is contacted with sulfuric acid to remove the less stable hydrocarbon molecules.

acid wash color--an indication of the presence of *olefins* and *polar compounds* in petroleum *solvents*. A sample of solvent is mixed with sulfuric acid and let stand until formation of an acid layer, the color of which is compared against color standards.

acrylic resin--any of a group of thermoplastic *resins* formed from the polymerization (see *polymer*) of acrylic acid, methacrylic acid, *esters* of these acids, or acrylonitrile. It is used in the manufacture of lightweight, weather-resistant, exceptionally clear plastics.

acute effect--toxic effect in mammals and aquatic life that rapidly follows exposure to a toxic substance. An acute effect is usually evident after a single oral intake, a single contact with the skin or eyes, or a single exposure to contaminated air lasting any period up to eight hours. Also known as acute toxicity.

acute toxicity--see *acute effect*.

additive--chemical substance added to a petroleum product to impart or improve certain properties. Common petroleum product additives are: *anti-foam agent, anti-icing additive, anti-wear additive, corrosion inhibitor, demulsifier, detergent, dispersant, emulsifier, EP additive, oiliness agent, oxidation inhibitor, pour point depressant, rust inhibitor, tackiness agent, viscosity index (V.I.) improver.*

adiabatic compression--compression of a gas without extraction of heat, resulting in increased temperature. The temperature developed in compression of a gas is an important factor in lubrication since oil deteriorates more rapidly at elevated temperatures. *Oxidation inhibitors* help prevent rapid lubricant breakdown under these conditions.

adjuvant--a part of a pesticide formulation that helps or adds to the action of the active ingredient. Petroleum products are sometimes used as adjuvants.

adsorption--adhesion of the molecules of gases, liquids, or dissolved substances to a solid surface, resulting in relatively high concentration of the molecules at the place of contact: e.g., the plating out of an *anti-wear additive* on metal surfaces. Also, any refining process in which a gas or a liquid is contacted with a solid, causing some compounds of the gas or liquid to adhere to the solid: e.g., contacting of lube oils with activated clay

to improve color. See *clay filtration.*

aerosol--a highly dispersed suspension of fine solid or liquid particles in a gas. Petroleum solvents are commonly used either as carriers or as vapor pressure depressants in packaged aerosol specialty products.

aftercooling--the process of cooling compressed gases under constant pressure after the final stage of compression. See *intercooling.*

afterrunning--the continued running of a spark-ignited engine after the ignition is turned off; also known as **dieseling**. There are two basic causes of afterrunning: **surface ignition** and **compression ignition**. In surface ignition, the surfaces of the combustion chamber remain hot enough to provide a source of ignition after the spark ignition is terminated. In compression ignition, the conditions of temperature, pressure, fuel composition, and engine idle speed allow ignition to continue.

arc hardening--increase in the consistency of a lubricating grease with storage time.

AGMA--American Gear Manufacturers Association, which as one of its activities establishes and promotes standards for gears and lubricants.

AGMA lubricant numbers-- AGMA specifications covering gear lubricants. The *viscosity* ranges of the AGMA numbers conform to the International Standards Organization (ISO) viscosity classification system (*see ISO viscosity classification system*). AGMA numbers and their viscosity ranges are as follows:

AGMA Lubricant Number	Corresponding ISO Grade	Viscosity Range cSt @ 40°C
1	46	41.4 - 50.6
2	68	61.2 - 74.8
3	100	90 - 110
4	150	135 - 165
5	220	198 - 242
6	320	288 - 352
6 Compounded	460	414 - 506
8 Compounded	680	612 - 748
8A Compounded	1000	900 - 1000
2 EP	68	61.2 - 74.8
4 EP	150	135 - 165
5 EP	220	198 - 242
6 EP	320	288 - 352
6 EP	460	414 - 506
8 EP	680	612 - 748

alcohol--any of a class of chemical compounds containing an hydroxyl OH group and having the general formula Cn $H_{2n}+_1$ OH: e.g., *methanol*, CH_3 OH: *ethanol*, C_2H_3 OH.

aliphatic hydrocarbon--hydrocarbon in which the carbon atoms are joined in open chains, rather than rings. See *hydrocarbon, normal paraffin*.

alkali--a hydroxide or carbonate of an alkali metal (e.g., lithium, sodium, potassium, etc), the aqueous solution of which is characteristically basic in chemical reactions. The term may be extended to apply to hydroxides and carbonates of

barium, calcium, magnesium, and the ammonium ion. See *base*.

alkyl--any of a series of monovalent *radicals* having the general formula $CnH_{2n}+_1$, derived from *aliphatic hydrocarbons* by the removal of a hydrogen atom: for example. CH_3. (methyl radical, from methane).

alkylate-- product of an *alkylation* process.

alkylated aromatic--benzene-derived *synthetic lubricant* base with good hydrolytic stability (resistance to chemical reaction with water) and good compatibility with *mineral oils*. Used in turbines, compressors, jet engines, and hydraulic power steering.

alkylation--in refining, the chemical reaction of a low-molecular-weight *olefin* with an *isoparaffin* to form a liquid product, **alkylate**, that has a high *octane number* and is used to improve the *antiknock* properties of gasoline. The reaction takes place in the presence of a strong acid *catalyst*, and at controlled temperature and pressure. Alkylation less commonly describes certain other reactions, such as that of an olefin with an *aromatic* hydrocarbon.

ambient--pertaining to any localized conditions, such as temperature, humidity, or atmospheric pressure, that may affect the operating characteristics of equipment or the performance of a petroleum product; e.g., a high ambient temperature may cause gasoline vapor lock in an automobile engine.

American Gear Manufacturers Association--see *AGMA*.

American National Standards Institute--see *ANSI*.

American Petroleum Institute--see *API*.

American Society for Testing and Materials--see *ASTM*.

American Society of Lubrication Engineers--see *ASLE*.

amphoteric--having the capacity to behave as either an *acid* or *base*; e.g., aluminum hydroxide. $Al(OH)_3$, which neutralizes acids to form aluminum salts and reacts with strong bases to form aluminates.

anesthetic effect--the loss of sensation with or without the loss of consciousness. It can be caused by the inhalation of volatile hydrocarbons.

anhydrous--devoid of water.

aniline point--lowest temperature at which a specified quantity of aniline (a benzene derivative) is soluble in a specified quantity of a petroleum product, as determined

by test method ASTM D 611; hence, an empirical measure of the solvent power of a hydrocarbon-- the lower the aniline point, the greater the solvency. Paraffinic hydrocarbons have higher aniline points than *aromatic* types. See *paraffin*.

anionic emulsified asphalt--see *emulsified anionic asphalt*.

ANSI (American National Standards Institute)--organization of industrial firms, trade associations, technical societies, consumer organizations, and government agencies, intended to establish definitions, terminologies, and symbols; improve methods of rating, testing, and analysis; coordinate national safety, engineering, and industrial standards; and represent U.S. interests in international standards work.

anti-foam agent--one of two types of *additives* used to reduce foaming in petroleum products; silicone oil to break up large surface bubbles, and various kinds of *polymers* that decrease the amount of small bubbles entrained in the oils. See *foaming, entrainment*.

anti-icing additive--substance added to gasoline to prevent ice formation on the throttle plate of a *carburetor*. Anti-icing *additives* are of two types: those that lower the freezing point of water, and those that alter the growth of ice crystals so that they remain small enough to be carried away in the air stream. See *carburetor icing*.

antiknock--resistance of a gasoline to *detonation* in a *combustion chamber*. See *knock, octane number*.

antiknock compounds--substances which raise the antiknock quality of a gasoline, as expressed by *octane number*. Historically, tetraethyl lead (see *lead alkyl*) has been the most common antiknock compound, but its use is being phased out under Environmental Protection Agency (*EPA*) regulations. Other additives of the oxygenated organic type--e.g., tertiary butyl alcohol (TBA) and methyl tertiary-butyl ether (MTBE)--are coming into increasing use as octane boosters in gasoline.

anti-oxidant--see *oxidation inhibitor*.

anti-seize compound--grease-like substance containing graphite or metallic solids, which is applied to threaded joints, particularly those subjected to high temperatures to facilitate separation when required.

anti-wear additive--*additive* in a lubricant that reduces friction and excessive wear. See *boundary lubrication*.

API (American Petroleum Institute)--trade association of petroleum producers, refiners, marketers, and transporters, organized for the advancement of the petroleum industry by conducting research, gathering and disseminating information, and maintaining cooperation between government and the industry on all matters of mutual interest. One API technical activity has been the establishment of *API Engine Service Categories* for lubricating oils.

API Engine Service Categories--gasoline and diesel engine oil quality levels established jointly by *API*, *SAE*, and *ASTM*, and sometimes called SAE or API/SAE categories; formerly called **API Engine Service Classifications**. API Service Categories are as follows:

Service Station Oils

SA straight mineral oil (no additives)

SB anti-oxidant, anti-scuff, but non-detergent

SC protection against high- and low-temperature deposits, wear, rust, and corrosion; meets car makers' warranty requirements for 1964-1967 models

SD improved protection over SC oils; meets warranty requirements for 1968-1971 models

SE improved protection over SD oils; meets warranty requirements for 1972-1980 models

SF improved anti-wear and anti-oxidation; meets warranty

requirements for 1980 and later models

Commercial Oils (Diesel Engine)

CA light-duty service; meets obsolete Military Specification MIL-L2104A

CB moderate-duty; meets MIL-L2104A. Supplement 1

CC moderate-to-severe duty; meets obsolete Military Specification MIL-L-2104B

CD severe duty; highest protection against high- and low-temperature deposits, wear, rust, corrosion; meets Military Specification MIL-L-2104C

API gravity--see *specific gravity*.

apparent viscosity--viscosity of a fluid that holds only for the *shear rate* (and temperature) at which the viscosity is determined. See *shear stress, Brookfield viscosity*.

aromatic--*unsaturated hydrocarbon* identified by one or more *benzene* rings or by chemical behavior similar to benzene. The benzene ring is characterized by three double bonds alternating with single bonds between carbon atoms (compare with *olefins*). Because of these multiple bonds, aromatics are usually more reactive and have higher solvency than *paraffins* and *naphthenes*. Aromatics readily undergo electrophylic substitution; that is, they react to add other active molecular groups, such as nitrates, *sulfonates*, etc. Aromatics are used extensively as *petrochemical* building blocks in the manufacture of pharmaceuticals, dyes, plastics, and many other chemicals.

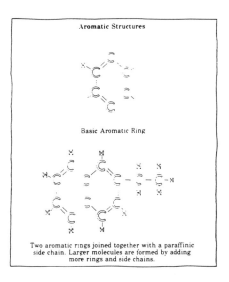

Aromatic Structures

Basic Aromatic Ring

Two aromatic rings joined together with a paraffinic side chain. Larger molecules are formed by adding more rings and side chains.

aryl--any organic group derived from an *aromatic* hydrocarbon by the removal of a hydrogen atom, for example C_6H_5. (phenyl *radical*, from *benzene*).

ash content--noncombustible residue of a lubricating oil or fuel, determined in accordance with test methods ASTM D 482 and D 874 (sulfated ash). Lubricating oil detergent additives contain metallic derivatives, such as barium, calcium, and magnesium sulfonates, that are common sources of ash. Ash deposits can impair engine efficiency and power. See *detergent*.

ashless dispersant--see *dispersant*.

askarel--generic term for a group of synthetic, fire-resistant, chlorinated *aromatic* hydrocarbons used as electrical insulating liquids. Gases produced in an askarel by arcing conditions consist predominantly of noncombustible hydrogen chloride, with lesser amounts of combustible gases. Manufacture of askarels has been discontinued in the U.S. because of their toxicity. See *PCB*.

ASLE (American Society of Lubrication Engineers)--organization intended to advance the knowledge and application of lubrication and related sciences.

asperities--microscopic projections on metal surfaces resulting from normal surface-finishing processes. Interference between opposing asperities in sliding or rolling applications is a source of friction, and can lead to metal welding and *scoring*. Ideally, the lubricating film between two moving surfaces should be thicker than the combined height of the opposing asperities. See *boundary lubrication, EP additive*.

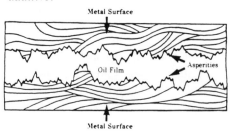

asphalt--brown-to-black, bituminous material (see *bitumen*) of high molecular weight, occurring naturally or as a residue from the distillation of crude oil; used as a bonding agent in road building, and in numerous industrial applications, including the manufacture of roofing. **Blown, or oxidized asphalt** is produced by blowing air through asphalt at high temperatures, producing a tougher, more durable asphalt. See also *asphalt cement, asphaltenes, emulsified anionic asphalt, emulsified cationic asphalt, penetration grading (asphalt), reclaimed asphalt pavement, recycling of asphalt paving, viscosity (asphalt), viscosity grading (asphalt)*.

asphalt cement (AC)--asphalt refined to meet specifications for paving and special purposes. Specifications are established by *ASTM* and the American Association of State Highway Transportation Officials (AASHTO).

asphaltenes--high-molecular-weight hydrocarbon components of *asphalt* and heavy residual stocks (see *bottoms*) that are soluble in carbon disulfide but not in paraffinic naphtha.

asphalt grading--see *penetration grading (asphalt), viscosity grading (asphalt)*.

asphaltic--containing significant amounts of *asphaltenes*.

Asphalt Institute--an international, non-profit association sponsored by members of the petroleum asphalt industry that serves both users and producers of asphaltic materials through programs of engineering service, research, and education.

asphalt paving--see *asphalt cement*.

asphalt recycling--see *recycling of asphalt paving*.

aspiration--drawing of air at atmospheric pressure into a *combustion chamber*; as opposed to supercharging or turbocharging. See *supercharger*.

ASTM (American Society for Testing and Materials)--organization devoted to "the promotion of knowledge of the materials of engineering, and the standardization of specifications and methods of testing." A preponderance of the data used to describe, identify, or specify petroleum products is determined in accordance with ASTM test methods.

ASTM scale (D 1500)--see *color scale*.

ATF--see *automatic transmission fluid*.

atmosphere--unit of *pressure* equal to 101.3 kilopascals (kPa) or 14.7 pounds per square inch (psi), or 760 mm (29.92 in) of mercury; standard atmospheric pressure at sea level.

atmospheric pollutants--see *pollutants*.

atmospheric pressure--see *pressure*.

atomization--the reduction of a liquid into fine particles or spray. Atomization is accomplished by the *carburetor* or the *fuel injection* system in *internal combustion engines*, and by special steam or air atomizers in furnaces and boilers.

austempering--see *quenching*.

auto-ignition temperature--lowest temperature at which a flammable gas or vaporized liquid will ignite in the absence of a spark or flame, as determined by test method ASTM D 2155; not to be confused with *flash point* or *fire point*, which is typically lower. Auto-ignition temperature is a critical factor in heat transfer oils and transformer oils, and in solvents used in high-temperature applications.

automatic transmission fluid (ATF)--fluid for automatic hydraulic transmissions in motor vehicles; also called **hydraulic transmission fluid**. Automatic transmission fluids must have a suitable *coefficient of friction*, good low-temperature viscosity, and anti-wear properties. Other necessary properties are: high *oxidation stability*, anticorrosion, anti-foaming, and compatibility with *synthetic rubber* seals. See *corrosion, foaming*.

automotive emissions--see *emissions (automotive)*.

aviation gasoline (avgas)--high-quality gasoline manufactured under stringent controls to meet the rigorous performance and safety requirements of piston-type aircraft engines. Volatility of aviation gasoline is closely controlled since, in most aircraft engines, excessive volatility can lead to *vapor lock*. Aviation gasolines are formulated to resist chemical degradation and to prevent fuel system corrosion. There are two basic grades of aviation gasolines (based on their *antiknock* value); 80 (80 lean/ 87 rich) and 100 (100 lean/ 130 rich). Aviation gasoline has different properties than *turbo fuel*, which fuels gas-turbine-powered aircraft. See *lean and rich octane numbers*.

azeotrope--liquid mixture of two or more components that boils at a temperature either higher or lower than the boiling point of any of the individual components. In refining, if the components of a solution are very close in boiling point and cannot be separated by conventional *distillation*, a substance can be added that forms an azeotrope with one component, modifying its boiling point and making it separable by distillation.

bactericide--additive included in the formulations of water-mixed *cutting fluids* to inhibit the growth of bacteria promoted by the presence of water, thus preventing the unpleasant odors that can result from bacterial action.

barrel--standard unit of measurement in the petroleum industry, equivalent to 42 standard U.S. gallons.

base--any of a broad class of compounds, including *alkalis*, that react with *acids* to form salts, plus water. Also known as hydroxides. Hydroxides ionize in solution to form hydroxyl ions (OH-); the higher the concentration of these ions, the stronger the base. Bases are used extensively in petroleum refining in *caustic washing* of *process streams* to remove acidic impurities, and are components in certain *additives* that neutralize weak acids formed during *oxidation*.

base coat--see *launching lubricant*.

base number--see *neutralization number*.

base stock--a primary refined petroleum fraction, usually a lube oil, into which *additives* and other oils are blended to produce finished products. See *distillation*.

basin--trough-like geological area, the former bed of an ancient sea. Because basins consist of sedimentary rock and have contours that provide traps for petroleum, they are considered good propects for exploration.

bearing--basic machine component designed to reduce friction between moving parts and to support moving loads. There are two main types of bearings: (1) **rolling contact bearings** (commonly ball or roller), and (2) **sliding (plain) bearings**, either plain journal (a metal jacket fully or partially enclosing a rotating inner shaft) or pad-type bearings, for linear motion. Rolling-contact bearings are more effective in reducing friction. With few exceptions, bearings require lubrication to reduce wear and extend bearing life.

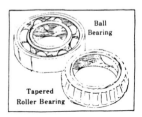

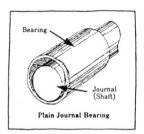

Plain Journal Bearing

benzene--aromatic hydrocarbon consisting of six carbon atoms and six hydrogen atoms arranged in a hexagonal ring structure. See *aromatic, hydrocarbon*. It is used extensively in the *petrochemical* industry as a chemical intermediate and reaction *diluent* and in some applications as a *solvent*. Benzene is a toxic substance, and proper safety precautions should be observed in handling it.

benzene

bevel gear--see *gears*.

bhp--brake horsepower, the effective or available power of an engine or turbine, measured at the output shaft. It is equivalent to the calculated horsepower, less the power lost in friction.

bitumen--any of various mixtures of viscous, brown-to-black hydrocarbons, such as *asphalt* and tar, together with any accompanying non-metallic derivatives, such as sulfur or nitrogen compounds; may occur naturally or may be obtained as residues from refining processes.

black oil--lubricant containing *asphaltic* materials, used in heavy-duty equipment applications, such as mining and quarrying, where extra adhesiveness is desired.

bleeding--the separation of liquid lubricant from a lubricating grease. See *syneresis*.

block grease--very firm grease manufactured in block form to be applied to certain large open plain bearings, generally operating at slow speeds and moderate temperatures. See *bearing*.

blocking point--lowest temperature at which waxed papers stick together, or block, sufficiently to injure the surface films and performance properties, as determined by test method ASTM D 1465.

block penetration--see *penetration (grease)*.

blow-by--in an *internal combustion engine*, seepage of fuel and gases past the piston rings and cylinder wall into the crankcase resulting in crankcase oil dilution and *sludge* formation. See *positive crankcase ventilation, dilution of engine oil*.

blown asphalt--see *asphalt*.

blown rapeseed oil--see *rapeseed oil*.

blowout--uncontrolled eruption of gas, oil, or other fluids from a well to the atmosphere.

blowout preventer--equipment installed at the *wellhead* to prevent the escape of pressure, or pressurized material, from the drill hole.

BMEP--brake mean effective pressure, the theoretical average pressure that would have to be imposed on the pistons of a frictionless engine (of the same dimensions and speed) to produce the same power output as the engine under consideration; a measure of how effectively an engine utilizes its piston displacement to do work.

boiling range--temperature spread between the initial boiling point and final boiling point. See *distillation test*.

bomb oxidation stability--measure of the *oxidation stability* of greases and lubricating oils in separate tests: ASTM D 942 (GREASE), astm d 2272 (oil), and ASTM D 2112 (insulating oils). In all tests, the sample is placed in a container, or bomb, which is then charged with oxygen and pressurized; a constant elevated temperature is maintained. ASTM 2272 utilizes a rotating bomb, which is placed in a heated bath; the test therefore is commonly called the **rotary bomb oxidation test**. Oxidation stability is expressed in terms of pressure drop in a given time period (D 942) or in terms of the time required to achieve a specified pressure drop (D 2272, D 2112).

borehole--the hole made by drilling, or boring, a well; also called **wellbore**. See *rotary drilling*.

bottled gas--a gas pressurized and stored in a transportable metal container. See *LPG*.

bottoms--in refining, the high-boiling residual liquid (also called **residuum**)--including such components as heavy fuels and asphaltic substances--that collects at the bottom of a distillation column, such as a pipe still. See *distillation, asphalt, fuel oil*.

boundary lubrication--form of lubrication between two rubbing surfaces without development of a full-fluid lubricating film. See *full-fluid-film lubrication, ZN/P curve*. Boundary lubrication can be made more effective by including additives in the lubricating oil that provide a stronger oil film, thus preventing excessive friction and possible *scoring*. There are varying degrees of boundary lubrication, depending on the severity of service. For mild conditions, *oiliness agents* may be used; these are *polar compounds* that have an exceptionally high affinity for metal surfaces. By plating out on these surfaces in a thin but durable film, oiliness agents prevent scoring under some conditions that are too severe for a *straight mineral oil*. *Compounded oils*, which are formulated with polar *fatty oils* are sometimes used for this purpose. *Anti-*

wear additives are commonly used in more severe boundary lubrication applications. High quality motor oils contain anti-wear additives to protect heavily loaded engine components, such as the valve train. The more severe cases of boundary lubrication are defined as extreme pressure conditions; they are met with lubricants containing *EP additives* that prevent sliding surfaces from fusing together at high local temperatures and pressures.

BR--see *polybutadiene rubber*.

brake horsepower--see *bhp*.

breakdown voltage--see *dielectric strength*.

bright stock--high viscosity oil, highly refined and dewaxed, produced from residual stocks, or *bottoms*; used for blending with lower viscosity oils.

British thermal unit--see *Btu*.

bromine index--number of milligrams of bromine that will react with 100 grams of a petroleum product (test method ASTM D 2710). Bromine index is essentially equivalent to *bromine number* 1000.

bromine number--number of grams of bromine that react with 100 grams of a sample of a petroleum *distillate*, giving an indication of its relative degree of reactivity, as determined by test method ASTM D 1159; it can be used as an indicator of the relative amount of *olefins* and *diolefins*, which are double-bonded straight-chain or cyclic hydrocarbons. See *hydrocarbon*.

Brookfield viscosity--*apparent viscosity* of an oil, as determined under test method ASTM D 2983. Since the apparent viscosity of a *non-Newtonian fluid* holds only for the *shear rate* (as well as temperature) at which it is determined, the Brookfield viscometer provides a known rate of shear by means of a spindle of specified configuration that rotates at a known constant speed in the fluid. The torque imposed by fluid friction can be converted to absolute viscosity units (centipoises) by a multiplication factor. See *viscosity, shear stress*. The viscosities of certain petroleum waxes and wax-*polymer* blends in the molten state can also be determined by the Brookfield test method ASTM D 2669.

BS&W--abbreviation of "bottoms sediment and water." The water and other extraneous material present in crude oil. Normally, the BS&W content must be quite low before the oil is accepted for pipeline delivery to a refinery.

Btu (British thermal unit)--quantity of heat required to raise the temperature of one pound of water one degree Fahrenheit, at 60°F and at a pressure of one *atmosphere*. See *energy*.

bulk appearance--visual appearance of grease when the undisturbed surface is viewed in an opaque container. Bulk appearance should be described in the following terms: **smooth**--a surface relatively free of irregularities; **rough**--a surface composed of many small irregularities; **grainy**--a surface composed of small granules or lumps of soap particles; **cracked**--showing surface cracks of appreciable number and magnitude; **bleeding**--showing free oil on the surface of the grease (or in the cracks of a cracked grease). See *texture*.

bulk delivery--large quantity of unpackaged petroleum product delivered directly from a tank truck, tank car, or barge into a consumer's storage tank.

bulk modulus--measure of a fluid's resistance to compressibility; the reciprocal of compressibility.

bulk odor--odor of vapor emanating from bulk liquid quantities of a petroleum product; also referred to as **impact odor**. The odor remaining after the product has evaporated is called **residual odor**.

Buna-N--see *nitrile rubber*.

Bunker C fuel oil--see *fuel oil*.

butadiene rubber--see *polybutadiene rubber*.

butane--gaseous paraffinic hydrocarbon (C_4H_{10}), usually a mixture of iso- and normal butane (see *isomer, normal paraffin*); also called, along with *propane*, liquefied petroleum gas (*LPG*).

normal butane

butylene--any of three isomeric (see *isomer*) flammable, gaseous hydrocarbons of the molecular structure C_4H_8; commonly derived from hydrocarbon *cracking*.

butyl rubber (IIR)--*synthetic rubber*, produced by copolymerization of isobutylene with isoprene or butadiene (see *polymer*). It is resistant to weather and heat, has low air-permeability and low resiliency; used in the manufacture of cable insulation, tubeless tire innerliners and other applications requiring good weather resistance and air retention.

°C (Celsius)--see *temperature scales*.

calcium soap grease--see *grease*.

calorie--term applicable either to the **gram calorie** or the **kilocalorie**. The gram calorie is defined as the amount of heat required at a pressure of one *atmosphere* to raise the temperature of one gram of water one degree Celsius at 15°C. The kilocalorie is the unit used to express the energy value of food; it is defined as the amountof heat required at a pressure of one atmosphere to raise the temperature of one kilogram of water one degree Celsius; it is equal to 1000 gram calories. See *energy*.

calorific value--see *heat of combustion*.

carbonizable substances--petroleum components which can be detected in *white oil, petrolatum,* and *paraffin wax* when any of these products is mixed with concentrated sulfuric acid, causing the acid to discolor, as outlined in test method ASTM D 565 (white oil and petrolatum) or ASTM D 612 (paraffin wax).

carbon monoxide (CO)--colorless, odorless, poisonous gas formed by the incomplete combustion of any carbonaceous material (e.g., gasoline, wood, coal). CO is the most widely distributed and most commonly occurring air pollutant, with motor vehicles being the primary source of man-made emissions, although emission controls are reducing the automobile's contribution. It is estimated that more than 90% of atmospheric CO comes from natural sources, such as decaying organic matter. See *catalytic converter, emissions (automotive), pollutants*.

carbon residue--percent of coked material remaining after a sample of oil has been exposed to high temperatures under test method ASTM D 189 (**Conradson**) or D 524 (**Ramsbottom**); hence, a measure of coke-forming tendencies. Results should be interpreted cautiously, as there may be little similarity between test conditions and actual service conditions.

carbon type analysis--empirical analysis of rubber process oil composition that expresses the percentages of carbon atoms in aromatic, naphthenic, and paraffinic components, respectively. See *rubber oil, aromatic, naphthene, paraffin*.

carbonyl--the divalent *radical* CO, which occurs in various organic substances, such as organic acids; also, any metal compound containing this radical. Carbonyls are highly reactive and considered to be catalyst poisons when present in solvents used as reaction *diluents* in *polyolefin* plastics manufacture.

carburetor--device in an internal combustion engine that atomizes and mixes fuel with air in the proper proportion for efficient combustion at all engine speeds, and controls the engine's power output by throttling, or metering, the air-fuel mixture admitted to the cylinders. The automobile carburetor is a complex mechanism designed to compensate for many variables over a wide range of speeds and loads. Intake air is drawn through the **venturi**, a constricted throat in the air passage that causes a pressure reduction in the air stream, which draws fuel from the carburetor bowl through either the **main jet** or the **idle jet**.

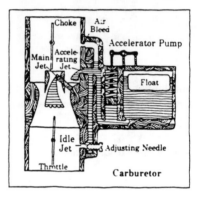

The fuel is atomized by the high-velocity air, and the resulting air-fuel mixture is piped through the intake manifold to the indiviudal cylinders, where it is burned. A **throttle plate** between the venturi and the cylinders controls power and speed by controlling the volume of air-fuel mixture reaching the cylinders. In most carburetors, closing of this (venturi) throttle valve shuts down the main jet and activates the idle jet, which provides the fuel-rich mixture that idling requires. An **accelerator pump** in the carburetor provides momentary fuel enrichment when the accelerator pedal is depressed rapidly, to compensate for the sudden influx of air. During cold starting, a **choke** (or butterfly valve) restricts airflow to the carburetor, thus enriching the mixture for faster starting. The choke on most automotive engine carburetors is operated automatically by a thermostatic spring, which opens the choke as the engine warms up. See *fuel injection, supercharger.*

carburetor icing--freezing of the moisture in humid air inside the *carburetor*, restricting air supply to the engine and causing it to stall. The air is brought to freezing by the chilling effect of vaporizing fuel. Carburetor icing is most likely to occur when the air temperature is between 3° and 13°C; if the *ambient* temperature were higher, the moisture would not provide sufficient moisture. Carburetor icing can be prevented by using a gasoline with an *anti-icing additive.*

carcinogen--cancer-causing substance.

carrier--a liquid, such as water, solvent, or oil, in which an active ingredient is dissolved or dispersed.

CAS (Chemical Abstract Service) Registry Numbers--identifying numbers assigned to chemical substances by the Chemical Abstract Service of the American Chemical Society and used by the Environmental Protection Agency

(*EPA*) to aid in registering chemicals under the federal Toxic Substances Control Act (TSCA) of 1976. CAS numbers are assigned to generic refinery *process streams*, such as *kerosene* and lube *base stocks*, that contain no *additives*. Petroleum products containing additives are termed "mixtures" by the TSCA and, as such, do not have CAS numbers. All chemical substances used in such mixtures are assigned CAS numbers and must be listed with the EPA by the refiner or the additive supplier.

casing--steel pipe placed in a *borehole* as drilling progresses, to prevent the wall of the hole from caving in. See *rotary drilling.*

casinghead gasoline--see *natural gasoline.*

catalyst--substance that causes or speeds up a chemical reaction without itself undergoing an associated change; catalysts are important in a number of refining processes.

catalytic converter--an emissions control device, incorporated into an automobile's exhaust system, containing catalysts--such as platinum, palladium, or rhodium--that reduce the levels of *hydrocarbons* (HC), *carbon monoxide* (CO), and--in more recent designs--*nitrogen oxides* (NOx) emitted to the air. In the catalytic converter, HC and CO are oxidized to form carbon dioxide (CO_2), and NOx are reduced to nitrogen and oxygen. Three-way catalytic converters that control all three substances require associated electronic controls for precise regulation of oxygen levels in the exhaust gas. Catalytic converters are also effective in removing *PNA* (polynuclear aromatic) hydrocarbons. Cars equipped with catalytic converters require unleaded gasoline, since the lead in tetraethyl lead, an *antiknock compound*, is a catalyst "poison." See *emissions (automotive), hydrocarbon emissions, pollutants, lead alkyl*.

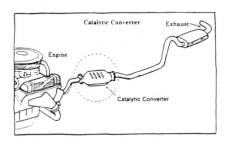

catalytic cracking--in refining, the breaking down at elevated temperatures of large, high-boiling hydrocarbon molecules into smaller molecules in the presence of a *catalyst*. The principal application of catalytic cracking is the production of high-octane gasoline, to supplement the gasoline produced by *distillation* and other processes. Catalytic cracking also produces heating oil components and hydrocarbon *feedstocks*, such as *propylene* and *butylene*, for *polymerization, alkylation*, and *petrochemical* operations.

cationic emulsified asphalt--see *emulsified cationic asphalt*.

caustic washing (scrubbing)--treatment of a petroleum liquid or gas with a caustic alkaline material (e.g., sodium hydroxide) to remove *hydrogen sulfide*, low-weight *mercaptans*, and other acidic impurities.

cavitation--formation of an air or vapor pocket (or bubble) due to lowering of pressure in a liquid, often as a result of a solid body, such as a propeller or piston, moving through the liquid; also, the pitting or wearing away of a solid surface as a result of the collapse of a vapor bubble. Cavitation can occur in a *hydraulic system* as a result of low fluid levels that draw air into the system, producing tiny bubbles that expand explosively at the pump outlet, causing metal erosion and eventual pump destruction. Cavitation can also result when reduced pressure in lubricating grease dispensing systems forms a void, or cavity, which impedes suction and prevents the flow of greases.

Celsius (°C)--see *temperature scales*.

centigrade--see *temperature scales*.

centipoise--see *viscosity*.

centistoke--see *viscosity*.

centralized lubrication--automatic dispensing of grease or oil from a reservoir to the lubricated parts of one or more machines. Flow is maintained by one or more pumps, and the amount of lubricant supplied to each point can be regulated by individual metering devices. Such a system provides *once-through lubrication*.

cetane--colorless liquid hydrocarbon. $C_{15}H_{34}$, used as a standard in determining *diesel fuel* ignition performance. See *cetane number*.

cetane improver--additive for raising the *cetane number* of a *diesel fuel*.

cetane index--an approximation of *cetane number* based on API gravity (see *specific gravity*) and mid-boiling point (see *distillation test*) of a *diesel fuel*. See *diesel index*.

cetane number--measure of the ignition quality of a *diesel fuel*, expressed as the percentage of cetane that must be mixed with liquid methylnaphthalene to produce the same ignition performance as the diesel fuel being rated, as determined by test method ASTM D 613. A high cetane number indicates shorter ignition lag and a cleaner burning fuel. See *cetane, cetane index, diesel index*.

channeling--formation of a channel in lubricating grease by a lubricated element, such as a gear or rolling contact bearing, leaving shoulders of grease that serve as a seal and reservoir. This phenomenon is usually desirable, although a channel that is too deep or permaent could cause lubrication failure.

Chemical Abstract Service Registry Numbers--see *CAS Registry Numbers*.

chlorine--see halogen.

Christmas tree--structure of valves, fittings, and pressure gauges at the top of a well to control the flow of oil and gas.

Christmas Tree

chronic effect--cumulative physiological damage resulting from prolonged exposure or series of exposures to a toxic substance. Also known as **chronic toxicity**.

chronic toxicity--see *chronic effect*.

circulating lubrication system--system in which oil is recirculated from a sump or tank to the lubricated parts, in most cases requiring a pump to maintain circulation. Circulating lubrication makes possible extended lubricant use, and usually requires a high-quality rust-and-oxidation-inhibited (*R&O*) oil.

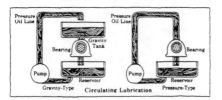

Gravity-Type   Circulating Lubrication   Pressure-Type

clay filtration--refining process using fuller's earth (activated clay) or bauxite to adsorb minute solids from lubricating oil, as well as remove traces of water, acids, and *polar compounds*. See *adsorption*.

clay/silica gel analysis--composition analysis test (ASTM D 2007) for determining weight percent of

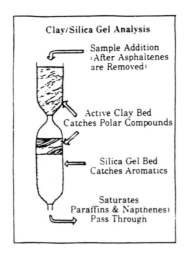

asphaltenes, saturated hydrocarbons, aromatics, and polar compounds in a petroleum product. The test material is mixed with pentane, asphaltenes are extracted as pentane insolubles, and the remainder of the sample is washed through a column. Active clay at the top of the column separates polar compounds, silica gel at the bottom separates aromatics, and saturates (*naphthenes* and *paraffins*) pass through with the pentane.

Cleveland Open Cup (COC)--test (ASTM D 92) for determining the *flash point* and *fire point* of all petroleum products except fuel oil and products with flash points below 79oC (175oF). The oil sample is heated in a precisely specified brass cup containing a thermometer. At specified intervals a small flame is passed across the cup. The lowest temperature at which the vapors above the cup briefly ignite is the flash point; the temperature at which the vapors sustain combustion for at least five seconds is the fire point. See *Tag open cup*.

closed cup--method for determining the *flash point* of fuels, *solvents*, and *cutback asphalts*, utilizing a covered container in which the test sample is heated and periodically exposed to a small flame introduced through a shuttered opening. The lowest temperature at which the vapors above the sample briefly ignite is the flash point. See

Pensky-Martens closed tester, Tag closed tester.

cloud point--temperature at which a cloud or haze of wax crystals appears at the bottom of a sample of lubricating oil in a test jar, when cooled under conditions prescribed by test method ASTM D 2500. Cloud point is an indicator of the tendency of the oil to plug filters or small orifices at cold operating temperatures. It is very similar to *wax appearance point.*

CMS asphalt--see *emulsified cationic asphalt.*

coastal oil--common term for any predominately *naphthenic* crude derived from fields in the Texas Gulf Coast area.

coefficient of friction--see *friction.*

cohesion--molecular attraction causing substances to stick together; a factor in the resistance of a lubricant, especially a grease, to flow.

cold-end corrosion--corrosion due to the acid-forming condensation of sulfur trioxide (SO_3) on cool surfaces of a boiler, especially the cooler parts of the chimney and the air heater; also called **low-temperature corrosion**. It can be prevented or minimized by using resistant alloys, by operating at low excess air levels (which reduces SO_3 production), or by operating at higher stack temperatures.

cold-flow improver--additive to improve flow of *diesel fuel* in cold weather. In some instances, a cold-flow improver may improve operability by modifying the size and structure of the wax crystals that precipitate out of the fuel at low temperatures, permitting their passage through the fuel filter. In most cases, the additive depresses the *pour point*, which delays agglomeration of the wax crystals, but usually has no significant effect on diesel engine performance. A preferred means of improving cold flow is to blend kerosene with the diesel fuel, which lowers the *wax appearance point* by about 1°C (2°F) for each 10% increment of kerosene added.

cold sett grease--see *sett grease.*

colloid--suspension of finely divided particles, 5 to 5000 angstroms in size, in a gas or liquid, that do not settle and are not easily filtered. Colloids are usually ionically stabilized by some form of surface charge on the particles to reduce the tendency to agglomerate. A lubricating grease is a colloidal system, in which metallic soaps or other thickening agents are dispersed in, and give structure to, the liquid lubricant.

color scale--standardized range of colors against which the colors of petroleum products may be compared. There are a number of widely used systems of color scales, including: **ASTM scale** (test method ASTM D 1500), the most common scale, used extensively for industrial and process oils; **Tag-Robinson colorimeter**, used with solvents, waxes, industrial and process oils; **Saybolt chrometer** (ASTM D 156), used with white oils, naphthas, waxes, fuels, kerosene, solvents; **Lovibond tintometer**, for USP petrolatums, sulfonates, chemicals; and the **Platinum-Cobalt (APHA) system** (ASTM D 1209), for lacquer solvents, diluents, petrochemicals. These scales serve primarily as indicators of product uniformity and freedom from contamination.

combustion--rapid oxidation of a fuel (burning). The products of an ideal combustion process are water (H_2O) and carbon dioxide (CO_2); if combustion is incomplete, some carbon is not fully oxidized, yielding *carbon monoxide* (CO). A *stoichiometric* combustible mixture contains the exact quantities of air (oxygen) and fuel required for complete combustion. For gasoline, this air-fuel ratio is about 15:1 by weight. If the fuel concentration is too rich or too lean relative to the oxygen in the mixture, combustion cannot take place. See *explosive limits*.

combustion chamber--in an *internal combustion engine*, the volume, bounded by the top of the piston and the inner surface of the cylinder head, in which the air-fuel charge ignites and burns. Valves and spark plugs are fitted into the combustion chamber.

commercial oils--see *API Engine Service Categories*.

complex soap--see *grease*.

compounded oil--mixture of a petroleum oil with animal or vegetable fat or oil. Compounded oils have a strong affinity for metal surfaces; they are particularly suitable for wet-steam conditions and for applications where *lubricity* and extra load-carrying ability are needed. They are not generally recommended where long-term *oxidation stability* is required.

compression ignition--see *after-running*.

compression-ignition engine--diesel engine. See *internal combustion engine*.

compression ratio--in an *internal combustion engine*, the ratio of the volume of the combustion space in the cylinder at the bottom of the piston stroke to the volume at the top of the stroke. High-compression-ratio gasoline engines require high octane fuels. Not to be confused with the *pressure ratio* of a compressor.

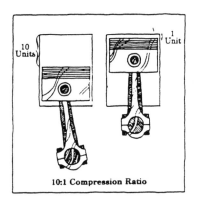

10:1 Compression Ratio

compressor--any of a wide variety of mechanisms designed to compress air or other gas to produce useful work. There are two basic types of compressors: **positive displacement** and **kinetic** (or dynamic). Positive displacement compressors increase pressure directly by reducing the volume of the chamber in which the gas is confined. Kinetic compressors are designed to impart velocity to the gas by means of a bladed rotor, then convert velocity energy to pressure by diverting the gas stream through stationary blades that change its direction of flow.

concrete form coating--an oil, wax, or grease applied to wooden or metal concrete forms to keep the hardened concrete from adhering to the forms. The liquid materials are also called **form oil**.

condensate--in refining, the liquid produced when hydrocarbon vapors are cooled. In oil and gas production, the term applies to hydrocarbons that exist in gaseous form under *reservoir* conditions, but condense to a liquid when brought to the surface.

congealing point--the temperature at which molten wax ceases to flow, as measured by test method ASTM D 938; of importance where storage or application temperature is a critical factor.

coning oil--lubricant, containing *emulsifiers* and anti-static agents, applied to synthetic-fiber yarn to reduce snagging and pulling as the yarn is run off a cone, and to facilitate further processing. See *fiber lubricant*.

Conradson carbon residue--see *carbon residue*.

conservation--see *energy conservation*.

consistency (grease)--a basic property describing the softness or hardness of a *grease*, i.e., the degree to which a grease resists deformation under the application of force. Consistency is measured by means of a cone penetration test. See *penetration (grease)*. The consistency of a grease depends on the viscosity of the base oil and the type and proportion of the thickener. It may also be affected by recent agitation; to take this phenomenon into consideration, a grease may be subjected to working (a standard churning process) prior to measuring its penetration value. See *NLGI consistency grades*.

copolymer--see *polymer*.

copper strip corrosion--the tendency of a petroleum product to corrode cuprous metals, as determined by test method ASTM D 130; the corrosion stains on a test copper strip are matched against standardized corroded strips.

corrosion--chemical attack on a metal or other solid by contaminants in a lubricant. Common corrosive contaminants are: (1) water, which causes rust, and (2) acids, which may form as oxidation products in a deteriorating oil, or may be introduced into the oil as combustion by-products in piston engines.

corrosion inhibitor--*additive* for protecting lubricated metal surfaces against chemical attack by water or other contaminants. There are several types of corrosion inhibitors. *Polar compounds* wet the metal surface preferentially, protecting it with a film of oil. Other compounds may absorb water by incorporating it in a water-in-oil emulsion so that only the oil touch-

es the metal surface. Another type of corrosion inhibitor combines chemically with the metal to present a non-reactive surface. See *rust inhibitor*.

cp (centipoise)--see *viscosity*.

CPSC (Consumer Products Safety Commission)--federal government commission that administers legislation on consumer product safety, such as the Consumer Product Safety Act, the Federal Hazardous Substances Act, the Flammable Fabrics Act, the Poison Prevention Packaging Act, and the Refrigeration Safety Act.

CR--see *neoprene rubber*.

cracking--petroleum refining process in which large-molecule liquid hydrocarbons are converted to small-molecule, lower-boiling liquids or gases; the liquids leave the reaction vessel as unfinished *gasoline, kerosene,* and *gas oils*. At the same time, certain unstable, more reactive molecules combine into larger molecules to form tar or coke *bottoms*. The cracking reaction may be carried out under heat and pressure alone (*thermal cracking*), or in the presence of a catalyst (*catalytic cracking*).

crankcase oil--see *engine oil*.

CRS asphalt--see *emulsified cationic asphalt*.

crude oil--complex, naturally occurring fluid mixture of petroleum hydrocarbons, yellow to black in color, and also containing small amounts of oxygen, nitrogen, and sulfur derivatives and other impurities. Crude oil was formed by the action of bacteria, heat, and pressure on ancient plant and animal remains, and is usually found in layers of porous rock such as *limestone* or *sandstone*, capped by an impervious layer of shale or clay that traps the oil (see *reservoir*). Crude oil varies in appearance and hydrocarbon composition depending on the locality where it occurs, some crudes being predominately naphthenic, some paraffinic, and others asphaltic. Crude is refined to yield petroleum products. See *distillation, hydrocarbon, sour crude, sweet crude, asphalt, naphthene, paraffin.*

CSS asphalt--see *emulsified cationic asphalt.*

cSt (centistoke)--see *viscosity.*

cut--segregated part, or fraction, separated from crude in the distillation process. See *distillation.*

cutback asphalt--a solution of *asphalt cement* and a petroleum *diluent.* Upon exposure to the atmosphere, the diluent evaporates, leaving the asphalt cement to perform its function. There are three grades: **RC asphalt**--rapid curing cutback asphalt, composed of asphalt cement and a *naphtha*-type diluent of high volatility; **MC asphalt**--medium-curing cutback asphalt, composed of asphalt cement and a *kerosene*-type diluent of medium volatility; **SC asphalt**--slow-curing cutback asphalt, composed of asphalt cement and a low-volatility oil. For industrial uses, oxidized asphalts can be blended with a petroleum diluent to meet the specific requirements of coatings, mastics, etc. (see *asphalt*)

cutting fluid--fluid, usually of petroleum origin, for cooling and lubricating the tool and work in machining and grinding. Some fluids are fortified with *EP additives* to facilitate cutting of hard metals, to improve finishes, and to lengthen tool life. Some cutting fluids are transparent to provide a better view of the work. Soluble cutting oils are emulsifiable with water to improve cooling. Since the resulting *emulsions* are subject to bacterial action and the development of odors, soluble cutting fluids may contain *bactericides.*

cyclic hydrocarbon--*hydrocarbon* in which the carbon atoms are joined in rings.

cycloalkane--see *naphthene.*

cycloparaffin--see *naphthene.*

cylinder oil--lubricant for independently lubricated cylinders, such as those of steam engines and air compressors; also for lubrication of valves and other elements in the cylinder area. Steam cylinder oils are available in a range of grades with high viscosities to compensate for the thinning effect of high temperatures; of these, the heavier grades are formulated for super-heated and high-pressure steam, and the less heavy grades for wet, saturated, or low-pressure steam. Some grades are compounded for service in excessive moisture; see *compounded oil*. Cylinder oils lubricate on a once-through basis (see *once-through lubrication*).

deasphalting--refining step for removal of *asphaltic* compounds from heavy lubricating oils. Liquid propane, liquid butane, or a mix-ture of the two is used to dilute the oil and precipitate the asphalt.

demerit rating--arbitrary graduated numerical rating sometimes used in evaluating engine deposit levels following testing of an engine oil's *detergent-dispersant* characteristics. On a scale of 0-10, the higher the number, the heavier the deposits. A more commonly used method of evaluating engine cleanliness is *merit rating*. See *engine deposits*.

demulsibility--ability of an oil to separate from water, as determined by test method ASTM D 1401 or D 2711. Demulsibility is an impor-tant consideration in lubricant maintenance in many *circulating lubrication systems*.

demulsifier--*additive* that promotes oil-water separation in lubricants that are exposed to water or steam. See *demulsibility*.

denaturing oil--unpalatable oil, commonly *kerosene* or No. 2 *heat-ing oil*, required to be added to food substances condemned by the U.S. Department of Agriculture, to ensure that these substances will not be sold as food or consumed as such.

density--see *specific gravity*.

de-oiling--removal of oil from petroleum wax; a refinery process usually involving filtering or press-ing a chilled mixture of *slack wax* and a solvent that is miscible in the oil, to lower the oil content of the wax.

depletion allowance--a reduction in U.S. taxes on producers of miner-als, including petroleum, to com-pensate for the exhaustion of an irreplaceable capital asset.

deposits--see *engine deposits*.

dermatitis--inflammation of the skin; can be caused by contact with many commercial substances, including petroleum products. Oil and grease in contact with the skin can result in plugging of sweat glands and hair follicles or defatting of the skin, which can lead to dermatitis. Dermatitis can be prevented in such cases by avoiding contact with the causative substances, or, if contact occurs, by promptly washing the skin with soap, water, and a soft skin brush. Clothes soaked with the substance should be removed.

detergent--important component of engine oils that helps control *varnish*, ring zone deposits, and rust by keeping insoluble particles in colloidal suspension (see *colloid*) and in some cases, by neutralizing acids. A detergent is usually a metallic (commonly barium, calcium, or magnesium) compound, such as a sulfonate, phosphonate, thiophosphonate, phenate, or salicylate. Because of its metallic composition, a detergent leaves a slight ash when the oil is burned. A detergent is normally used in conjunction with a *dispersant*. See *ash content*.

detergent-dispersant--engine oil *additive* that is a combination of a *detergent* and a *dispersant*; important in preventing the formation of *sludge* and other engine deposits.

detonation--see *knock*.

dewaxing--removal of *paraffin wax* from lubricating oils to improve low temperature properties, especially to lower the *cloud point* and *pour point*.

dibasic acid ester (diester)--*synthetic lubricant* base; an organic ester, formed by reacting a

dicarboxylic acid and an *alcohol*; properties include a high *viscosity index* (V.I.) and low *volatility*. With the addition of specific additives, it may be used as a lubricant in compressors, hydraulic systems, and internal combustion engines.

dielectric--non-conductor of electricity, such as *insulating oil* for transformers. See *power factor*.

dielectric loss--see *power factor*.

dielectric strength (breakdown voltage)--minimum voltage required to produce an electric arc through an oil sample, as measured by test method ASTM D 877; hence, an indication of the insulating (arc preventive) properties of a transformer oil. A low dielectric strength may indicate contamination, especially by water. See *insulating oil, power factor*.

diesel engine--see *internal combustion engine*.

diesel fuel--that portion of crude oil that distills out within the temperature range of approximately 200°C (392°F) to 370°C (698°F), which is higher than the boiling range of gasoline. See *distillation*. Diesel fuel is ignited in an *internal combustion engine* cylinder by the heat of air under high compression--in contrast to motor gasoline, which is ignited by an electrical spark. Because of the mode of ignition, a high *cetane number* is required in a good diesel fuel. Diesel fuel is close in boiling range and composition to the lighter *heating oils*. There are two grades of diesel fuel, established by the American Society for Testing and Materials (*ASTM*): **Diesel 1** and **Diesel 2**. Diesel 1 is a *kerosene*-type fuel, lighter, more volatile, and cleaner burning than Diesel 2, and is used in engine applications where there are frequent changes in speed and load. Diesel 2 is used in industrial and heavy mobile service.

diesel index--an approximation of the *cetane number* of a *distillate* fuel; the product of the API gravity (see *specific gravity*) and the *aniline point* (°F) divided by 100. Fuels of unusual composition may show erroneous cetane numbers by this method. See *cetane index*.

dieseling--see *afterrunning*.

diester--see *dibasic acid ester*.

diluent--a usually inert (unreactive) liquid or *solvent*, used to dilute, carry, or increase the bulk of some other substance. Petroleum oils and solvents are commonly used as diluents in such products as paints, pesticides, and additives. See *reaction diluent*.

dilution of engine oil--thinning of *engine oil* by seepage of fuel into the crankcase, as measured by test method ASTM D 322, which indicates the volume percentage of fuel in the sample. Dilution is detrimental to lubrication, and may indicate defective engine components--such as worn piston rings--or improper fuel system adjustment.

dimer--see *polymerization*.

diolefin--highly reactive straight-chain hydrocarbon with two double bonds between adjacent carbon atoms. See *olefin*.

dispersant--engine oil *additive* that helps prevent *sludge, varnish,* and other engine deposits by keeping particles suspended in a colloidal state (see *colloid*). Dispersants are normally used in conjunction with *detergents*. A dispersant is commonly distinguished from a detergent in that the former is non-metallic and, thus, does not leave an ash when the oil is burned; hence, the term **ashless dispersant**. Also, a dispersant can keep appreciably larger quantities of contaminants in suspension than a detergent.

dispersion--minute discrete particles suspended in a liquid, a gas, or a solid. Though it may have the general characteristics of a *colloid*, a dispersion is not necessarily a truly homogeneous mixture.

dissipation factor--in an electrical system with an applied alternating voltage, it is expressed as the tangent of the loss angle (δ) or the cotangent of the phase angle ($\propto$); a measure of the degree of electrical loss due to the imperfect nature of an insulating liquid surrounding the electrical system, as determined by test method ASTM D 924. Dissipation factor is related to *power factor*.

distillate--any of a wide range of petroleum products produced by distillation, as distinct from *bottoms*, cracked stock (see *cracking*), and *natural gas liquids*. In fuels, a term referring specifically to those products in the mid-boiling range, which include *kerosene, turbo fuel,* and *heating oil*--also called **middle distillates** and **distillate fuels**. In lubricating oils, a term applied to the various fractions separated under vacuum in a distillation tower for further processing (lube distillate). See *distillation*.

distillate fuels--see *distillate*.

distillation **(fractionation)**--the primary refining step, in which crude is separated into **fractions**, or components, in a **distillation tower**, or **pipe still**. Heat, usually applied at the bottom of the tower, causes the oil vapors to rise through progressively cooler levels of the tower, where they condense onto plates and are drawn off in order of their respective condensation temperatures, or boiling points --the lighter-weight, lower-boiling-point fractions, exiting higher in the tower. The primary fractions, from low to high boiling point, are: hydrocarbon gases (e.g., *ethane,*

propane); *naphtha* (e.g., *gasoline*); *kerosene, diesel fuel* (*heating oil*); and heavy *gas oil* for *cracking*.

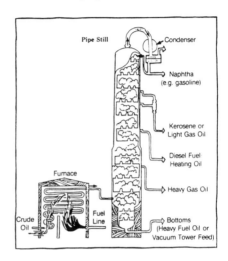

Heavy materials remaining at the bottom are called the *bottoms*, or residuum, and include such components as heavy fuel oil (see *fuel oil*) and asphaltic substances (see *asphalt*). Those fractions taken in liquid form from any level other than the very top or bottom are called *sidestream* products; a product, such as propane, removed in vapor form from the top of the distillation tower is called *overhead* product. Distillation may take place in two stages: first, the lighter fractions--gases, naphtha, and kerosene--are recovered at essentially atmospheric pressure; next, the reamining crude is distilled at reduced pressure in a **vacuum tower**, causing the heavy lube fractions to distill at much lower temperatures than possible at atmospheric pressure, thus permitting more lube oil to be distilled without the molecular cracking that can occur at excessively high temperatures.

distillation test--method for determining the full range of *volatility* characteristics of a hydrocarbon liquid by progressively boiling off (evaporating) a sample under controlled heating. **Initial boiling point** (IBP) is the fluid temperature at which the first drop falls into a graduated cylinder after being condensed in a condenser connected to a distillation flask. **Mid-boiling point** (MBP) is the temperature at which 50% of the fluid has collected in the cylinder. **Dry point** is the temperature at which the last drop of fluid disappears from the bottom of the distillation flask. **Final boiling point** (FBP) is the highest temperature observed. **Front-end volatility** and **tail-end volatility** are the amounts of test sample that evaporate, respectively, at the low and high temperature ranges. If the boiling range is small, the fluid is said to be **narrow cut**, that is, having components with similar volatilities; if the boiling range is wide, the fluid is termed **wide cut**. Distillation may be carried out by several ASTM test methods, including ASTM D 86, D 850, D 1078, and D 1160.

distillation tower--see *distillation*.

dN factor--also called **speed factor**, determined by multiplying the bore of a rolling-contact bearing (d), in millimeters, by the speed (N), in rpm. of the journal (the shaft or axle supported by the bearings); used in conjunction with operating temperature to help determine the appropriate *viscosity* of the bearing lubricating oil. See *bearing*.

dolomite--sedimentary rock similar to *limestone*, but rich in magnesium carbonate. Dolomite is sometimes a *reservoir* rock for petroleum.

drawing compound--in metal forming, a lubricant for the die or blank used to shape the metal; often contains *EP additives* to increase die life and to improve the surface finish of the metal being drawn.

drilling--see *rotary drilling*.

drilling fluid--also called **drilling mud**. See *mud*.

drilling mud--see *mud*.

dropping point--lowest temperature at which a grease is sufficiently fluid to drip, as determined by test method ASTM D 566 or D 2265; hence, an indication of whether a grease will flow from a bearing at operating temperatures. The test is of limited significance in predicting overall service performance.

dry point--see *distillation test.*

dumbbell blend--mixture of hydro-carbons, usually two components, that have markedly different volatilities, viscosities, or other properties. See *volatility, viscosity.*

dynamic demulsibility--test of water separation properties of an oil, involving continuous mixture of oil and water at elevated tempera-tures in an apparatus that simulates a lubricating oil circulating system. Samples are then drawn off both the top and bottom of the test appa-ratus. Ideally, the top sample should be 100% oil, and the bottom 100% water. Because of the sever-ity of the test conditions, separation is virtually never complete.

elasto-hydrodynamic (EHD) lubri-cation--lubrication phenomenon occurring during elastic deforma-tion of two non-conforming surfac-es under high load. A high load carried by a small area (as between

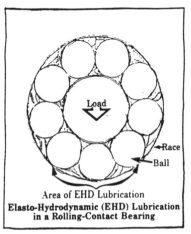

Area of EHD Lubrication
Elasto-Hydrodynamic (EHD) Lubrication in a Rolling-Contact Bearing

the ball and race of a rolling con-tact bearing) causes a temporary increase in lubricant viscosity as the lubricant is momentarily trapped between slightly deformed opposing surfaces.

elastomer--rubber or rubber-like material, both natural and synthetic, used in making a wide variety of products, such as ties, seals, hose, belting, and footwear. In oil seals, an elastomer's chemical composi-tion is a factor in determining its compatibility with a lubricant, par-ticularly a *synthetic lubricant*. See *natural rubber, synthetic rubber.*

electronic emission controls (EEC)--in automobiles, computer-ized engine operating controls that reduce automotive exhaust emis-sions of *carbon monoxide* (CO), *nitrogen oxides* (NOx), and *hydro-carbons* (HC), primarily by opti-mizing combustion efficiency. This is accomplished by automatic moni-toring and control of key engine functions and operating parameters, such as air-fuel ratio (see *combus-tion*), spark timing, and *exhaust gas recirculation*. See *emissions (auto-motive), hydrocarbon emissions.*

electrostatic precipitation--re-moval of particles suspended in a gas--as in a furnace flue--by elec-trostatic charging of the particles, and subsequent precipitation onto a collector in a strong electrical field. See *emissions (stationary source), particulates, pollutants.*

emission controls--see *catalytic converter, emissions (automotive), emissions (stationary source), electronic emission controls, exhaust gas recirculation, positive crankcase ventilation.*

emissions (automotive)--the three major pollutant emissions for which gasoline-powered vehicles are controlled are: unburned *hydrocarbons* (HC), *carbon monoxide* (CO), and *nitrogen oxides* (NOx). Diesel-powered vehicles primarily emit NOx and *particulates*. Motor vehicles contribute only a small percentage of total man-made emissions of other atmospheric pollutant, such as *sulfur oxides*. Evaporative HC emissions from the fuel tank and carburetor are adsorbed by activated carbon contained in a canister installed on the vehicle. *Blow-by* HC emissions from the crankcase are controlled by *positive crankcase ventilation* (PCV). Exhaust emissions of HC, CO, and NOx--the products of incomplete combustion--are controlled primarily by a *catalytic converter*, in conjunction with *exhaust gas recirculation* and increasingly sophisticated technology for improving combustion efficiency, including *electronic emission controls*. See *emissions (stationary source), hydrocarbon emissions, pollutants.*

emissions (stationary source)--atmospheric pollutants from fossil fuel combustion in furnaces and boilers. Stationary combustion sources contribute significantly to total man-made emissions of *sulfur oxides*--predominantly sulfur dioxide (SO_2), with some sulfur trioxide (SO_3)--*nitrogen oxides* (NOx), and *particulates*, but emit comparatively minor amounts of *carbon monoxide* (CO) and *hydrocarbons* (HC). Means of controlling stationary source emissions include: flue gas scrubbing with a chemical substance such as sodium hydroxide to remove sulfur, burning naturally low-sulfur fuels, lowering combustion temperatures to reduce NOx formation, and using *electrostatic precipitation* to reduce particulate emissions. See *emissions (automotive), pollutants.*

emulsified cationic asphalt (cationic emulsified asphalt)--emulsified *asphalt* in which the asphalt globules are electronegatively charged. Common grades of anionic asphalt are: **RS asphalt**--anionic rapid-setting emulsified asphalt; **CMS asphalt**--cationic medium-setting emulsified asphalt; **css asphalt**--cationic slow-setting emulsified asphalt. Emulsified cationic asphalt is superior to *emulsified anionic asphalt* in its ability to mix with wet stones, or aggregate, because its electropositive charge aids in rapidly replacing the water adhering to the stones.

emulsifier--*additive* that promotes the formation of a stable mixture, or *emulsion*, of oil and water. Common emulsifiers are: metallic soaps, certain animal and vegetable oils, and various *polar compounds* (having molecules that are water-soluble at one extremity of their structures and oil-soluble at the other).

emulsion--intimate mixture of oil and water, generally of a milky or cloudy appearance. Emulsions may be of two types: oil-in-water (where water is the continuous phase) and water-in-oil (where water is the discontinuous phase). Oil-in-water emulsions are used as *cutting fluids* because of the need for the cooling effect of the water. Water-in-oil emulsions are used where the oil, not the water, must contact a surface--as in *rust preventives*, non-flammable *hydraulic fluids*, and compounded steam *cylinder oils* (see *compounded oil*); such emulsions are sometimes referred to as **inverse emulsions**. Emulsions are produced by adding an *emulsifier*. Emulsibility is not a desirable characteristic in certain lubricating oils, such as crankcase or turbine oils, that must separate from water readily.   Unwanted emulsification can occur as a result of oxidation products--which are usually *polar compounds*--or other contaminants in the oil. See illustration of an oil-in-water emulsion at *polar compound*.

energy--capacity to do work. There are many forms of energy, any of which can be converted into any other form of energy.   To produce electrical power in a steam turbine-generator system, the chemical energy in coal is converted into heat energy, which (through steam) is converted to the mechanical energy of the turbine, and in turn, converted into electrical energy. Electrical energy may then be converted into the mechanical energy of a vacuum cleaner, the radiant and heat energy of a light bulb, the chemical energy of a charged battery, etc.   Conversion from one form of energy to another results in some energy being lost in the process (usually as heat). There are two kinds of mechanical energy: **kinetic energy**, imparted by virtue of a body's motion, and **potential energy**, imparted by virtue of a body's position (e.g., a coiled spring, or a stone on the edge of a cliff).   Solar (radiant) energy is the basis of all life through the process of photosynthesis, by which green plants convert solar energy into chemical energy. Nuclear energy is the result of the conversion of a small amount of the mass of an unstable (radioactive) atom into energy. The fundamental unit of energy in the *Systeme International* is the *joule*.   It can be expressed in other energy units, such as the *calorie, British thermal unit (Btu), kilowatt-hour*, etc. by use of appropriate conversion factors.

energy conservation--employment of less energy to accomplish the same amount of useful work; also, the reduction or elimination of any energy-consuming activity. Energy conservation is a vital goal in U.S. and world efforts to adjust to the declining availability of conventional oil and gas resources. Conservation will extend the time available to make the transition to alternative energy sources, such as solar energy and *synthetic oil and gas*, and will also reduce the economic hardship imposed by rising energy prices.

engine deposits--hard or persistent accumulations of *sludge, varnish*, and carbonaceous residues due to *blow-by* of unburned and partially burned (partially oxidized) fuel, or from partial breakdown of the crankcase lubricant. Water from condensation of combustion products, carbon, residues from fuel or lubricating oil additives, dust, and metal particles also contribute. Engine deposits can impair engine performance and damage engine components by causing valve and ring sticking, clogging of the oil screen and oil passages, and excessive wear of pistons and cylinders. Hot, glowing deposits in the *combustion chamber* can also cause *pre-ignition* of the air-fuel mix. Engine deposits are increased by short trips in cold weather, high-temperature operation, heavy loads (such as pulling a trailer), and over-extended oil drain intervals.

engine oil (crankcase oil, motor oil)--oil carried in the crankcase, sump, or oil pan of a reciprocating *internal combustion engine* to lubricate all major engine parts; also used in reciprocating compressors and in steam engines of crankcase design. In automotive applications, it is the function of the engine oil not only to lubricate, but to cool hot engine parts, keep the engine free of rust and deposits (see *engine deposits*), and seal the rings and valves against leakage of combustion gases. Oil-feed to the engine parts is generally under pressure developed by a gear pump (forced feed). The oil circulates through passages formed by tubing and drilling (rifling) through the engine parts, and through an oil filter to remove metallic contaminants and other foreign particles. In some engines, lubrication may also be accomplished in part by splashing resulting from the rotation of the crankshaft in the oil in the sump. Modern engine oils are formulated with *additives* to improve performance. Additive content in a single-viscosity-grade oil is typically around 15 mass percent, and in a multi-grade oil, 20 percent or more. See *API Engine Service Categories, military specifications for engine oils, SAE viscosity grades*.

Engler viscosity--method for determining the *viscosity* of petroleum products; it is widely used in Europe, but has limited use in the U.S. The test method is similar to *Saybolt Universal viscosity*; viscosity values are reported as "Engler degrees."

enhanced recovery--in crude oil production, any method used to produce the oil remaining in a reservoir that has largely been depleted. See *secondary recovery, tertiary recovery, reservoir.*

entrainment--the state of a liquid or gas that is dispersed, but undissolved, in a liquid or gaseous medium.

Environmental Protection Agency--see *EPA.*

EPA (Environmental Protection Agency)--agency of the federal executive branch, established in 1970 to abate and control pollution through monitoring, regulation, and enforcement, and to coordinate and support environmental research.

EP additive--lubricant *additive* that prevents sliding metal surfaces from seizing under conditions of extreme pressure (EP). At the high local temperatures associated with metal-to-metal contact, an EP additive combines chemically with the metal to form a surface film that prevents the welding of opposing *asperities*, and the consequent *scoring* that is destructive to sliding surfaces under high loads. Reactive compounds of sulfur, chlorine, or phosphorus are used to form these inorganic films.

EPDM rubber--see *ethylene-propylene rubber.*

EP oil--lubricating oil formulated to withstand extreme pressure (EP) operating conditions. See *EP additive.*

ester--chemical compound formed by the reaction of an organic or inorganic *acid* with an *alcohol* or with another organic compound containing the hydroxyl (-OH) *radical*. The reaction involves replacement of the hydrogen of the acid with a hydrocarbon group. The name of an ester indicates its derivation: e.g., the ester resulting from the reaction of ethyl alcohol and acetic acid is called ethyl acetate. Esters have important uses in the formulation of some petroleum additives and *synthetic lubricants.* See *dibasic acid ester, phosphate ester.*

ethane--gaseous paraffinic hydrocarbon (C_2H_6) present in natural gas and petroleum: used as a fuel, and as a *feedstock* in *petrochemical* manufacture. See *natural gas, distillation.*

$$H-\overset{\overset{\displaystyle H}{|}}{\underset{\underset{\displaystyle H}{|}}{C}}-\overset{\overset{\displaystyle H}{|}}{\underset{\underset{\displaystyle H}{|}}{C}}-H$$

ethane

ethanol--also known as ethyl alcohol (C_2H_5OH). Obtained principally from the fermentation of grains or blackstrap molasses; also obtained from *ethylene*, by absorption in sulfuric acid and hydrolyzing with water. Widely used as an industrial solvent, *extraction* medium, chemical intermediate, and in many proprietary products; a component of *gasohol*.

ethylene--flammable gas (C_2H_4) derived from natural gas and petroleum; the lowest molecular weieght member of the generic family of *olefins*. Ethylene is widely used as a *feedstock* in the manufacture of *petrochemicals*, including *polyethylene* and other plastics.

```
H   H
|   |
C = C
|   |
H   H
```
ethylene

ethylene-propylene rubber (EPM and EPDM)--*synthetic rubber*: EPM is a *polymer* of *ethylene* and *propylene*; EPDM is a polymer of ethylene and propylene with a small amount of a third *monomer* usually a *diolefin*) to permit vulcanization with sulfur.   EPM and EPDM possess excellent resistance to ozone, sunlight, and weathering, have good flexibility at low temperatures, and good electrical insulation properties. Used in the manufacture of tires, hoses, auto parts, coated fabrics, and electrical insulation.

evaporation--conversion of a liquid into a vapor; also, a test procedure that yields data on the *volatility* of a petroleum product. Evaporation testing of solvents may be performed in accordance with the Federation of Societies for Paint Technology Method II.   In this method, a small sample of product is applied by hypodermic syringe to a filter paper on a sensitive spring balance, then allowed to evaporate under controlled conditions of temperature, relative humidity, air movement, etc. The loss of sample weight is plotted with respect to time. See *distillation test*.

exhaust gas recirculation (EGR)-- system designed to reduce automotive exhaust emissions of *nitrogen oxides* (NOx). The system routes exhaust gases into the *carburetor* or intake manifold; the gases dilute the air-fuel mixture (see *combustion*) which lowers peak combustion temperatures, thus reducing the tendency for NOx to form.

explosive limits (flammability limits)--upper and lower limits of petroleum vapor concentration in air outside of which combustion will not occur. As a general rule, below one volume percent concentration in air (lower explosive limit) the mixture is too lean to support combustion; above six volume percent (upper explosive limit), the mixture is too rich to burn.   See *combustion*.

extender--material added to a formulation to improve quality and processability, or to reduce costs by substituting for a more expensive material (e.g., a low-cost solvent partially replacing a higher-cost solvent; a petroleum oil added to a rubber formulation).

extraction--use of a solvent to remove edible and commercial oils from seeds (e.g., soybeans), or oils and fats from meat scraps; also, the removal of reactive components from lube distillates (see *solvent extraction*) or other refinery *process streams*.

extreme pressure (EP) additive--see *EP additive*.

°F (Fahrenheit)--see *temperature scales*.

Falex test--a method for determining the extreme-pressure (EP) or anti-wear properties of oils and greases. Vee blocks (with a large "V"-shaped notch) are placed on opposite sides of a rotating steel shaft, and the apparatus is immersed in a bath of the test lubricant. Load is automatically in-

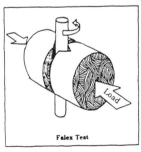

Falex Test

creased until seizure occurs. Measurable wear scars are formed on the blocks.

false brinelling--see *fretting*.

fatty acid--any monobasic (one displaceable hydrogen atom per molecule) organic acid having the general formula C_nH_{2n+1} COOH. Fatty acids derived from natural fats and oils are used to make soaps used in the manufacture of greases and other lubricants. See *grease*.

fatty oil--organic oil of animal or vegetable origin; can be added to petroleum oils to increase load-carrying ability, or oiliness. See *compounded oil, saponification number*.

FDA (Food and Drug Administration)--agency administered under the U.S. Department of Health and Human Services (formerly Health, Education and Welfare) "to enforce the Federal Food, Drug, and Cosmetic Act and thereby carry out the purpose of Congress to insure that foods are safe, pure, and wholesome, and made under sanitary conditions; drugs and therapeutic devices are safe and effective for their intended uses; cosmetics are safe and prepared from appropriate ingredients; and that all of these products are honestly and informatively labeled and packaged."

feedstock--any material to be processed; e.g., *gas oil* for *cracking*, *ethylene* for *petrochemical* manufacture.

FIA analysis--see *fluorescent indicator adsorption*.

fiber--in *grease*, form in which soap thickeners occur. On the average, soap fibers are about 20 times as long as they are thick; most are microscopic, so that the grease appears smooth.

fiber lubricant--an oil containing *emulsifiers* and anti-static agents, applied to synthetic fibers to lubricate them during processing into yarn; also called **spin finish**. See *coning oil*.

film strength--see *lubricity*.

final boiling point--see *distillation test*.

fingerprint   neutralizer--*polar compound* in some *rust preventives* that places a barrier between the metal   surface   and   perspiration deposited during handling of metal parts. In this way, corrosive activity of the salts and acids in perspiration is suppressed.

fire   point--temperature   at   which the vapor concentration of a combustible liquid is sufficient to sustain combustion, as determined by test method ASTM D 92, *Cleveland Open Cup*. See *flash point*.

fire-resistant fluid--lubricant used especially in high-temperature or hazardous hydraulic applications. Three common types of fire-resistant fluids are: 1) water-petroleum oil *emulsions*, in which the water prevents burning of the petroleum constituent; 2) water-glycol fluids; and 3) non-aqueous fluids of low volatility, such as *phosphate esters*, *silicones*,   and   halogenated   (see *halogen*) hydrocarbon-type fluids. See *flame propagation, synthetic lubricant*.

flame propagation--self-sustaining burning of fuel after heat of combustion has been reached.   Many fire-resistant   hydraulic   fluids-- though they can be made to burn if subjected to sufficiently intense heat--do not generate sufficient heat of combustion of themselves to continue burning once the external source of heat is removed.   See *fire-resistant fluids*.

flammability limits--see *explosive limits*.

flash point--lowest temperature at which the vapor of a combustible liquid can be made to ignite momentarily in air, as distinct from *fire point*. Flash point is an important indicator of the fire and explosion hazards associated with a petroleum product.   There are a number of ASTM tests for flash point, e.g., *Cleveland open cup*,

Pensky-Martens closed tester, Tag closed tester, Tag open cup.

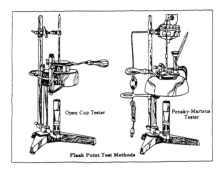

Open Cup Tester   Pensky-Martens Tester

Flash Point Test Methods

floc point--temperature at which waxy materials in a lubricating oil separate from a mixture of oil and Freon* R-12 refrigerant, giving a cloudy appearance to the mixture; also called **Freon floc point**. Generally used to evaluate the tendency of *refrigeration oils* to plug expansion valves or capillaries in refrigerant systems. Not to be confused with *cloud point*, the temperature at which wax precipitates from an undiluted oil.
*Registered trademark of E. I. DuPont de Nemours, Inc.

fluid friction--see *friction.*

fluorescent indicator adsorption (FIA)--method of measuring the relative concentration of *saturated hydrocarbons*, *aromatics*, and *olefins* in a petroleum product (usually a *solvent* or light *distillate*), as determined by test method ASTM D 1319. The sample is passed through a column where it reacts with three dyes, each sensitive to one of the components. The relative concentration of the component is indicated by the length of the respective dyed zones, viewed under ultraviolet light, which brings out the coloration of the dyes.

flux (flux oil)--a relatively non-volatile fraction of petroleum used as a *diluent* to soften asphalt to a desired consistency; also, a *base stock* for the manufacture of roofing asphalts.

foaming--occurrence of frothy mixture of air and a petroleum product (e.g., lubricant, fuel oil) that can reduce the effectiveness of the product, and cause sluggish hydraulic operation, air binding of oil pumps, and overflow of tanks or sumps. Foaming can result from excessive agitation, improper fluid levels, air leaks, *cavitation*, or contamination with water or other foreign materials. Foaming can be inhibited with an *anti-foam agent*. The foaming characteristics of a lubricating oil can be determined by blowing air through a sample at a specified temperature and measuring the volume of foam, as described in test method ASTM D 892.

fogging oil--light *mineral seal oil* injected in small quantities into a gas transmission line to settle dust or to seal joints by soaking the fiber or jute gasket materials. Also, oil used to generate smoke or fog to obscure the movement of troops and naval vessels, or as a carrier for insecticides applied to large outdoor areas.

follower plate--heavy disc in a grease container which rests on the surface of the grease, and assists its downward movement toward a dispensing pump located at the bottom of the container.

food   additive--non-nutritional substance added directly or indirectly to food during processing or packaging.   Petroleum food additives are usually *refined waxes* or *white oils* that meet applicable *FDA* standards.   Applications include direct additives, such as coatings for fresh fruits and vegetables, and indirect additives, such as impregnating oils for fruit and vegetable wrappers,   dough   divider   oils, defoamers for yeast and beet sugar manufacture, release and polishing agents in confectionery manufacture, and rust preventives for meat processing equipment.

Food and Drug Administration-- see *FDA*.

form oil--see *concrete form coating*.

fossil fuel--any fuel, such as crude oil and coal, derived from remains of ancient organisms that have been transformed over the ages by heat, pressure, and chemical action.

four-ball method--either of two lubricant test procedures, the Four-Ball Wear Method (ASTM D 2266) and Four-Ball EP (extreme pressure) Method (ASTM D 2596), based on the same principle. Three steel balls are clamped together to form a cradle upon which a fourth ball rotates on a vertical axis. The balls are immersed in the lubricant under investigation. The **Four-Ball Wear Method** is used to determine the anti-wear properties of lubricants operating under *boundary lubrication* conditions. The test is carried out at a specified speed, temperature, and load. At the end of a specified test time, the average diameter of the wear scars on the three lower balls is reported. The **Four-Ball EP Method** is designed to   evaluate   performance   under much   higher   unit   loads.   The loading is increased at specified intervals until the rotating ball seizes and welds to the other balls.

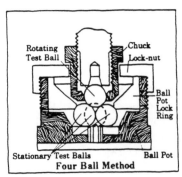

Four Ball Method

At the end of each interval the average scar diameter is recorded. Two values are generally reported-- *load wear index* (formerly **mean Hertz load**) and *weld point*.

four-square gear oil tester--device consisting of two automotive drive-axle systems to test the load-carrying capacity of hypoid gear (see *gear*) lubricants.

four-stroke-cycle--see *internal combustion engine*.

fraction, fractionation--see *cut, distillation*.

freezing point--a specific temperature that can be defined in two ways, depending on the ASTM test used. In ASTM D 1015, which measures the freezing point of high-purity petroleum products (such as *nitration-grade* toluene), freezing point is the temperature at which a liquid solidifies. In ASTM D 2386, which measures the freezing point of aviation fuel, freezing point is that temperature at which hydrocarbon crystals formed on cooling disappear when the temperature of the fuel is allowed to rise.

Freon floc point--see *floc point*.

fretting--form of wear resulting from small-amplitude oscillations or vibrations that cause the removal of very finely divided particles from rubbing surfaces (e.g., the vibra-tions imposed on the wheel bearings of an automobile when transported by rail car). With ferrous metals the wear particles oxidize to a reddish, abrasive iron oxide, which has the appearance of rust or corrosion, and is therefore sometimes called **fretting corrosion;** other terms applied to this phenomenon are **false Brinelling** (localized fretting involving the rolling elements of a bearing) and **friction oxidation.**

fretting corrosion--see *fretting*.

friction--resistance to the motion of one surface over another. The amount of friction is dependent on the smoothness of the contacting surfaces, as well as the force with which they are pressed together. Friction between unlubricated solid bodies is independent of speed and area. The **coefficient of friction** is obtained by dividing the force required to move one body over a horizontal surface at constant speed by the weight of the body; e.g., if a force of 4 kilograms is required to move a body weighing 10 kilograms, the coefficient of friction if 4/10, or 0.4. Coefficients of rolling friction (e.g., the motion of a tire or ball bearing) are much less than coefficients of sliding friction (back and forth) motion over two flat surfaces). Sliding friction is thus more wasteful of energy and can cause more wear. **Fluid friction** occurs between the molecules of a gas or liquid in motion, and is expressed as *shear stress*. Unlike solid friction, fluid friction varies with speed and area. In general, lubrication is the substitution of low fluid friction in place of high solid-to-solid friction. See *tribology*.

friction oxidation--see *fretting*.

front-end volatility--see *distillation test*.

fuel-economy oil--engine oil specially formulated to increase fuel efficiency. A fuel-economy oil works by reducing the *friction* between moving engine parts that wastefully consumes fuel energy. There are two known means of accomplishing this goal: 1) by reducing the *viscosity* of the oil to decrease fluid friction and 2) by using friction-reducing additives in the oil to prevent metal-to-metal contact, or rubbing friction, between surface *asperities*.

fuel injection--method of introducing fuel under pressure through a small nozzle into the intake system of cylinders of an engine. Fuel injection is essential to the diesel cycle, and an alternative to conventional carburetion in the gasoline engine. In some designs, each cylinder has a cam-operated injector, which is a plunger pump that delivers precisely metered quantities of fuel at precise intervals. The fuel is injected in a minutely divided spray at high discharge

pressures.   The amount of the charge is controlled by the throttle pedal.   A combination of fuel injection and carburetion is used in advanced emission-control systems, involving fuel injection into the throttle body of the carburetor. Fuel injection offers certain advantages over carburetion, including: more balanced fuel distribution in the cylinders for improved combustion, more positive delivery of fuel to the cylinder (hence, easier starting and faster acceleration), and higher power output because of improved *volumetric efficiency*. See *carburetor*.

fuel oil--term encompassing a broad range of *distillate* and residual fuels identified by ASTM grades 1 through 6. Grade No. 1, a *kerosene*-type fuel, is a light distillate fuel that has the lowest boiling range.   No. 2 fuel oil, popularly called *heating oil*, has a higher boiling range and is commonly used in home heating. It is comparable in boiling range to *diesel fuel*. Grades 4, 5, and 6 are called **heavy fuel oils** (HFO), or **residual fuel oils**; they are composed largely of heavy pipe still *bottoms*.   Because of their high viscosity, No. 5 and No. 6 fuel oils require preheating to facilitate pumping and burning.   No. 6 fuel oil is also called **Bunker C fuel oil**.

fuel pump--mechanism for delivering fuel from the tank to the *carburetor* of a gasoline engine, to the fuel injectors of a diesel engine, or to the fuel atomizers of an oil-fired boiler.

full-fluid-film lubrication--presence of a continuous lubricating film sufficient to completely separate two surfaces, as distinct from *boundary lubrication*. Full-fluid-film lubrication is normally **hydrodynamic lubrication**, whereby the oil adheres to the moving part and is drawn into the area between the sliding surfaces, where it forms a pressure, or hydrodynamic, wedge.

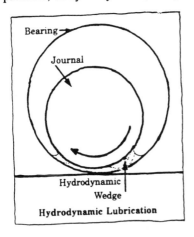

Hydrodynamic Lubrication

See *ZN/P curve*. A less common form of full-fluid-lubrication is **hydrostatic lubrication**, wherein the oil is supplied to the bearing area under sufficient external pressure to separate the sliding surfaces.

fully refined wax--see *refined wax*.

furfural--colorless liquid, C_4H_3OCHO, employed in petro-leum refining as a solvent to extract *mercaptans*, *polar compounds*, *aromatics*, and other impurities from oils and waxes. Also used in the manufacture of dyes and plastics.

Furol viscosity--viscosity of a petroleum oil measured with a Saybolt Furol viscometer; see *viscosity*.

FZG four-square gear oil test--test used in developing industrial gear lubricants to meet equipment manufacturers' specifications. The FZG test equipment consists of two gear sets, arranged in a four-square configuration, driven by an electric motor. The test gear set is run in the lubricant at gradually increased load stages until failure, which is the point at which a 10 milligram weight loss by the gear set is recorded. Also called **Niemann four-square gear oil test**.

gas blanket--atmosphere of inert gas (usually nitrogen or carbon dioxide) lying above a lubricant in a tank and preventing contact with air. In the absence of such a covering, the lubricant would be subject to oxidation. Gas blankets are commonly used with *heat transfer fluids* and electrical *insulating oils*.

gas engine--*internal combustion engine*, either two- or four-stroke cycle, powered by natural gas or *LPG*. Commonly used to drive compressors on gas pipelines, utilizing as fuel a portion of the gas being compressed.

gasohol--blend of 10 volume percent anhydrous *ethanol* (ethyl alcohol) and 90 volume percent unleaded gasoline.

gas oil--liquid petroleum *distillate*, higher boiling than *naphtha*; initial boiling point may be a low as 204°C (400°F). Gas oil is called light or heavy, depending on its final boiling point. It is used in blending *fuel oil* and as refinery *feedstock* in *cracking* operations.

gasoline--blend of light hydrocarbon fractions of relatively high *antiknock* value. Finished motor and *aviation gasolines* may consist of the following components: straight-run *naphthas*, obtained by the primary *distillation* of crude oil; *natural gasoline*, which is "stripped," or condensed, out of natural gas; cracked naphthas; reformed naphthas; and alkylate. (See *alkylation, catalytic cracking, reforming*.) A high-quality gasoline has the following properties: 1) pro- per *volatility* to ensure easy starting and rapid warm-up; 2) clean-burning characteristics to prevent harmful *engine deposits*; 3) *additives* to prevent rust, *oxidation*, and *carburetor icing*; 4) sufficiently high *octane number* to prevent engine knock.

gas turbine--see *internal combustion engine, turbine*.

gauge pressure--see *pressure*.

gear--machine part which transmits motion and force from one rotary shaft to another by means of successively engaging projections, called teeth. The smaller gear of a pair is called the **pinion**; the larger, the gear. When the pinion is on the driving shaft, the gear set acts as a speed reducer; when the gear drives, the set acts as a speed multiplier. The basic gear type is the **spur gear**, or **straight-tooth gear**, with teeth cut parallel to the gear axis. Spur gears transmit power in applications utilizing parallel shafts. In this type of gear, the teeth mesh along their full length, creating a sudden shift in load from one tooth to the next, with consequent noise and vibration. This problem is overcome by the **helical gear**, which has teeth cut at an angle to the center of rotation, so that the load is transferred progressively along the length of the tooth from one edge of the gear to the other. When the

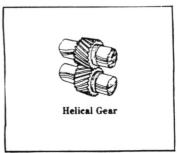

Helical Gear

shafts are not parallel, the most common gear type used is the **bevel gear**, with teeth cut on a sloping gear face, rather than parallel to the shaft. The **spiral bevel gear** has teeth cut at an angle to the plane of rotation, which, like the helical gear, reduces vibration and noise. A **hypoid gear** resembles a spiral bevel gear, except that the pinion is offset so that its axis does not intersect the gear axis; it is widely used in automobiles between the engine driveshaft and the rear axle. Offset of the axes of hypoid gears introduces additional sliding between the teeth, which, when combined with high loads, requires a high-quality EP oil. A **worm gear** consists of a spirally grooved screw moving against a toothed wheel; in this type of gear, where the load is transmitted across sliding, rather than rolling, surfaces, *compounded*

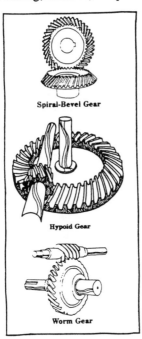

Spiral-Bevel Gear

Hypoid Gear

Worm Gear

oils or *EP oils* are usually necessary to maintain effective lubrication.

gear oil (automotive)--long-life oil of relatively high viscosity for the lubrication of rear axles and some manual transmissions. Most final drives and many accessories in agricultural and construction equipment also require gear oils. Straight (non-additive) mineral gear oils are suitable for most spiral-bevel rear axles (see *gear*) and for some manual transmissions. Use of such oils is declining, however, in favor of EP (extreme pressure) gear oils (see *EP oil*) suitable both for hypoid gears (see *gear*) and for all straight mineral oil applications. An EP gear oil is also appropriate for off-highway and other automotive applications for which the lubricant must meet the requirements of Military Specification MIL-L-2105C.

gear oil (industrial)--high-quality oil with good *oxidation stability*, rust protection, and resistance to *foaming*, for service in gear housings and enclosed chain drives. A *turbine oil* or R&O oil is the usual gear oil recommendation. Specially formulated industrial EP gear oils (see *EP oil*) are used where highly loaded gear sets or excessive sliding action (as in worm gears) is encountered. See *gear*.

gear shield--highly adhesive lubricant of heavy consistency, formulated with *asphaltic* compounds or *polymers* for protection of exposed gears and wire rope in circumstances where the lubricant cannot readily be replenished. Many gear shield lubricants must be softened with heat or cut back with solvents before they can be applied.

general purpose oils--see *once-through lubrication.*

gilsonite--a naturally occurring *asphalt* mined from rock fissures. It is hard and brittle, and has a high melting point. It is used in the manufacture of *rust preventives,* paints, sealants, and lacquers.

gloss--property of wax determinable by measuring light reflected from a wax-treated paper surface. Gloss stability is evaluated after a sample of treated paper has been held at an elevated temperature for a specified period.

gram calorie--see *calorie.*

graphite--a soft form of elemental carbon, gray to black, in color. It occurs naturally or is synthesized from coal or other carbon sources. It is used in the manufacture of paints, lead pencils, crucibles, and electrodes, and is also widely used as a lubricant, either alone or added to conventional lubricants.

grease--mixture of a fluid lubricant (usually a petroleum oil) and a thickener (usually a soap) dispersed in the oil. Because greases do not flow readily, they are used where extended lubrication is required and where oil would not be retained. The thickener may play as important a role as the oil in lubrication. **Soap thickeners** are formed by reacting (saponifying) a metallic hydroxide, or *alkali,* with a fat, *fatty acid,* or *ester.* The type of soap used depends on the grease properties desired. **Calcium** (lime) **soap greases** are highly resistant to water, but unstable at high temperatures. **Sodium soap greases** are stable at high temperatures, but wash out in moist conditions. **Lithium soap greases** resist both heat and moisture. A mixed-base soap is a combination of soaps, offering some of the advantages of each type. A **complex soap** is formed by the reaction of an alkali with a high-molecular-weight fat or fatty acid to form a soap, and the simultaneous reaction of the alkali with a short-chain organic or inorganic acid to form a metallic salt (the complexing agent). Complexing agents usually increase the *dropping point* of grease. Lithium, calcium, and aluminum greases are common alkalis in complex-soap greases. **Non-soap thickeners,** such as clays, silica gels, carbon black, and various synthetic organic materials are also used in grease manufacture. A *multi-purpose grease* is designed to provide resistance to heat, as well as water, and may contain additives to increase load-carrying ability and inhibit rust. See *consistency (grease)*, *penetration (grease)*.

grease gun injury--see *high-pressure-injection injury.*

gum in gasoline--oily, viscous contaminant that may form due to oxidation during storage. Gum formation in gasoline can cause serious fuel system problems, such as carburetor malfunctioning and intake valve sticking. See *oxidation*. The amount of gum in motor gasoline, aviation gasoline, and aircraft turbine fuel can be determined by evaporating a measured sample by means of air or steam flow at controlled temperature, and weighing the residue, as described in test method ASTM D 381.

halogen--any of a group of five chemically related nonmetallic elements: chlorine, bromine, fluorine, iodine, and astatine. Chlorine compounds are used as *EP additives* in certain lubricating oils, and as constituents of certain *petrochemicals* (e.g., vinyl chloride, chlorinated waxes). Chlorine and fluorine compounds are also used in some *synthetic lubricants*.

heat of combustion--measure of the available energy content of a fuel, under controlled conditions specified by test method ASTM D 240 or D 2382. Heat of combustion is determined by burning a small quantity of a fuel in an oxygen bomb calorimeter and measur-ing the heat absorbed by a specified quantity of water within the calorimeter. Heat of combustion is expressed either as *calories* per gram or *British thermal units* per pound. Also called **thermal value, heating value, calorific value.**

heat transfer fluid--circulating medium (often a petroleum oil) that absorbs heat in one part of a system (e.g., a solar heating system or a remote oil-fired system) and releases it to another part of the system. Heat transfer fluids require high resistance to cracking (molecular breakdown) when used in systems with fluid temperatures above 260°C (500°F). Systems can be either closed or open to the atmosphere. To prevent oxidation in a closed system inert gas is sometimes used in the expansion tank (or reservoir) to exclude air (oxygen). See *gas blanket*. If the system is open and the fluid is exposed simultaneously to air and to temperatures above 66°C (150°F), the fluid must also have good *oxidation stability*, since a protective gas blanket cannot be contained.

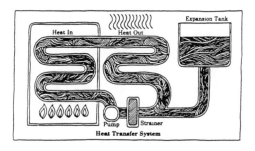

Heat Transfer System

heat treating oil--see *quenching oil*.

heating oil--see *fuel oil*.

heating value--see *heat of combustion*.

heavy crude naphtha--see *naphtha*.

heavy ends--highest boiling portion in a distilled petroleum fraction or finished product. In motor gasoline, the heavy ends do not fully volatilize until the engine has warmed. See *light ends*.

heavy fuel oil--see *fuel oil*.

helical gear--see *gear*.

heptane--liquid paraffinic (see *paraffin*) hydrocarbon containing seven carbon atoms in the molecule, which may be straight-chain (normal) or branched-chain (iso). Heptane can be used in place of *hexane* where a less volatile solvent is desired, as in the manufacture of certain adhesives and lacquers, and in *extraction* of edible and commercial oils. Heptane is blended with *isooctane* to create a standard reference fuel in laboratory determinations of *octane number*.

hexane--highly volatile paraffinic (see *paraffin*) hydrocarbon containing six carbon atoms in the molecule; it may also contain six-carbon *isoparaffins*. Widely used as a solvent in adhesive and rubber solvent formulations, and in the *extraction* of a variety of edible and commercial oils. Hexane is a neurotoxin and must be handled with adequate precautions.

high-pressure-injection injury--injury caused by the accidental injection of grease or oil under pressure through the skin and into the underlying tissue; also called a **grease gun injury**. Such an injury requires immediate medical attention.

homogenization--intimate mixing of a lubricating grease or an *emulsion* by intensive shearing action to obtain more uniform dispersion of the components.

horsepower--unit of power equal to 33,000 foot-pounds per minute, equivalent to 745.7 watts.

hot-dip rust preventive--petroleum-base *rust preventive*, consisting of a blend of oil, wax, or asphalt, and rust-inhibiting additives, that must be melted before application.

humidity--water vapor in the atmosphere. **Absolute humidity** is the amount of water vapor in a given quantity of air; it is not a function of temperature. **Relative humidity** is a ratio of actual atmospheric moisture to the maximum amount of moisture that could be carried at a given temperature, assuming constant atmospheric pressure. The higher the temperature--other factors remaining constant--the lower the relative humidity (i.e., the drier the air).

hydrated soap--grease thickener that has water incorporated into its structure to improve structural stability of the grease. See *grease*.

hydraulic fluid--fluid serving as the power transmission medium in a *hydraulic system*. The most commonly used fluids are petroleum oils, *synthetic lubricants*, oil-water emulsions, and water-glycol mixtures. The principal requirements of a premium hydraulic fluid are proper *viscosity*, high *viscosity index*, anti-wear protection (if needed), good *oxidation stability*, adequate *pour point*, good *demulsibility*, rust inhibition (see *rust inhibitor*), resistance to *foaming*, and compatibility with seal materials. Anti-wear oils are frequently used in compact, high-pressure, and high-capacity pumps that require extra lubrication protection. Certain *synthetic lubricants* and water-containing fluids are used where fire resistance is needed. See *fire-resistant fluids*.

hydraulic system--system designed to transmit power through a liquid medium, permitting multiplication of force in accordance with Pascal's law, which states that "a pressure exerted on a confined liquid is transmitted undiminished in all directions and acts with equal force on al equal areas." Hydraulic systems have six basic components: 1) a reservoir to hold the fluid supply; 2) a fluid to transmit the power; 3) a pump to move the fluid; 4) a valve to regulate pressure; 5) a directional valve to control the flow, and 6) a working component--such as a cylinder and piston or a shaft rotated by pressurized fluid--to turn hydraulic power into mechanical motion. Hydraulic systems offer several advantages over mechanical systems: they eliminate complicated mechanisms such as cams, gears, and levers; are less subject to wear; are usually more easily adjusted for control of speed and force; are easily adaptable to both rotary and linear transmission of power; and can transmit power over long distances and in any direction with small losses.

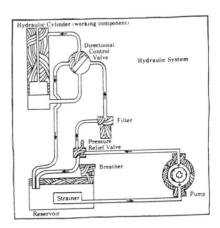

hydraulic transmission fluid--see *automatic transmission fluid*.

hydraulic turbine--see *turbine*.

hydrocarbon--chemical compound of hydrogen and carbon; also called an *organic compound*. Hydrogen and carbon atoms can be combined in virtually countless ways to make a diversity of products.   Carbon atoms form the skeleton of the hydrocarbon molecule, and may be arranged in chains (aliphatic) or rings (cyclic).   There are three principal types of hydrocarbons that occur naturally in petroleum: *paraffins*,   *naphthenes*,   and *aromatics*, each with distinctive properties. Paraffins are aliphatic, the others cyclic.   Paraffins and naphthenes are saturated; that is, they have a full complement of hydrogen atoms and, thus, only single bonds between carbon atoms. Aromatics are unsaturated, and have as part of their molecular structure at least one *benzene* ring, i.e., six carbon atoms in a ring configuration with alternating single and double bonds.   Because of these double bonds, aromatics are usually more reactive than paraffins and naphthenes, and are thus prime starting materials for chemical synthesis. Other types of hydrocarbons are formed during the petroleum refining process. Important among these are *olefins* and *acetylenes*. Olefins are unsaturated hydrocarbons with at least one double bond in the molecular structure, which may be in either an open chain or ring configuration; olefins   are   highly   reactive.

Acetylenes are also unsaturated and contain at least one triple bond in the molecule.   See *saturated hydrocarbons, unsaturated hydrocarbons*.

hydrocarbon (HC) emissions--substances considered to be atmospheric pollutants because the more reactive *hydrocarbons* (e.g., *aromatics*) undergo a photochemical reaction with *nitrogen oxides (NOx)* to form oxidants, components of smog that can cause eye irritation and respiratory problems.   Motor vehicles account for about one-third of man-made hydrocarbon emissions, although automotive emission controls are reducing this amount. The greatest portion of total atmospheric hydrocarbons is from natural sources, such as pine trees. See *catalytic converter, emissions (automotive), pollutants*.

hydrocracking--refining process in which middle and heavy *distillate* fractions are cracked (broken into smaller molecules) in the presence of hydrogen at high pressure and moderate temperature to produce high-octane gasoline, turbine fuel components, and middle distillates with good flow characteristics and *cetane* ratings.   The process is a combination of *hydrogenation* and *cracking*.

hydrodynamic   lubrication--see *full-fluid-film lubrication*.

Hydrofining®--form of *hydrogen treating* in which refinery distillate, lube, and wax streams are treated with hydrogen at elevated temperatures and moderate pressures in the presence of a *catalyst*, to improve color and stability, and reduce sulfur content. The patented process was developed by Exxon in 1951.

hydroforming--a dehydrogenation process in which *naphthas* are passed over a solid *catalyst* at elevated temperatures and moderate pressures in the presence of hydrogen to form high-octane motor gasoline, high-grade aviation gasoline, or aromatic solvents. The process is a net producer of hydrogen.

hydrogenation--in refining, the chemical addition of hydrogen to a *hydrocarbon* in the presence of a *catalyst*; a severe form of *hydrogen treating*. Hydrogenation may be either destructive or nondestructive. In the former case, hydrocarbon chains are ruptured (cracked) and hydrogen is added where the breaks have occurred. In the latter, hydrogen is added to a molecule that is unsaturated (see *unsaturated hydrocarbon*) with respect to hydrogen. In either case, the resulting molecules are highly stable. Temperature and pressures in the hydrogenation process are usually greater than in *Hydrofining®*.

hydrogen sulfide (H₂S)--gaseous compound of sulfur and hydrogen commonly found in crude oil; it is extremely poisonous, corrosive, and foul-smelling.

hydrogen treating--refining process in which *hydrocarbons* are treated with hydrogen in the presence of a *catalyst* at relatively low temperatures to remove *mercaptans* and other sulfur compounds, and improve color and stability. See *Hydrofining®*.

hydrolytic stability--ability of additives and certain *synthetic lubricants* to resist chemical decomposition (hydrolysis) in the presence of water.

hydrometer--see *specific gravity*.

hydrophilic--also **hygroscopic**, having an affinity for water. Some *polar compounds* are simultaneously hydrophilic and oil soluble.

hydrophobic--the opposite of *hydrophilic*.

hydrostatic lubrication--see *full-fluid-film lubrication*.

hygroscopic--see *hydrophilic*.

hypoid gear--see *gear*.

IBP--initial boiling point; see *distillation test*.

IFT--see *interfacial tension*.

immiscible--incapable of being mixed without separation of phases.

Water and petroleum oil are immiscible under most conditions, although they can be made miscible with the addition of an *emulsifier*. See *miscible*.

impact odor--see *bulk odor*.

induction period--the time period in an *oxidation* test during which oxidation proceeds at a constant and relatively low rate. It ends at a point at which the rate of oxidation increases sharply.

industrial asphalt--oxidized asphalt (see *asphalt*) used in the manufacture of roofing, asphaltic paints, *mastics*, and adhesives for laminating paper and foil. Industrial asphalt is generally harder than *asphalt cement*, which is used for paving.

industrial lubricant--any petroleum or synthetic-base fluid (see *synthetic lubricant*) or grease commonly used in lubricating industrial equipment, such as gears, turbines, compressors.

inhibitor--additive that improves the performance of a petroleum product through the control of undesirable chemical reactions. See *corrosion inhibitor, oxidation inhibitor, rust inhibitor*.

initial boiling point (IBP)--see *distillation test*.

inorganic   compound--chemical compound, usually mineral, that does not include *hydrocarbons* and their derivatives. However, some relatively simple carbon compounds, such as carbon dioxide, metallic carbonates, and carbon disulfide are regarded as inorganic compounds.

insoluble resins--see *insolubles*.

insolubles--test for contaminants in used lubricating oils, under conditions prescribed by test method ASTM D 893. The oil is first diluted with *pentane*, causing the oil to lose its solvency for certain oxidation *resins*, and also causing the precipitation of such extraneous materials as dirt, soot, and wear metals. These contaminants are called **pentane insolubles**. The pentane insolubles may then be treated with *toluene*, which dissolves the oxidation resins (benzene was formerly used). The remaining solids are called **toluene insolubles**. The difference in weight between the pentane insolubles and the toluene insolubles is called **insoluble resins**.

insulating oil--high-quality oxidation-resistant oil refined to give long service as a *dielectric* and coolant for transformers and other electrical equipment. Its most common application is as a **transformer oil**. An insulating oil must resist the effects of elevated temperatures, electrical stress, and contact with air, which can lead to sludge formation and loss of insulation properties. It must be kept dry, as water is detrimental to *dielectric strength*.

intercooling--cooling of a gas at constant pressure between stages in a compressor. It permits reduced work in the compression phase because cooler gas is more easily compressed. *Aftercooling* is the final cooling following the last compression stage.

interfacial tension (IFT)--the force required to rupture the interface between two liquid phases. The interfacial tension between water and a petroleum oil can be determined by measuring the force required to move a platinum ring upward through the interface, under conditions specified by test method ASTM D 971. Since the interface can be weakened by oxidation products in the oil, this measurement may be evidence of oil deterioration. The lower the surface tension below the original value, the greater the extent of oxidation. ASTM D 971 is not widely used with additive-containing oils, since additives may affect surface tension, thus reducing the reliability of the test as an indicator of oxidation.

internal combustion engine--heat engine driven directly by the expansion of combustion gases, rather than by an externally produced medium, such a steam. Basic versions of the internal combustion engine are: **gasoline engine** and **gas engine** (spark ignition), **diesel engine** (compression ignition), and **gas turbine** (continuous combustion). Diesel compression-ignition engines are more fuel-efficient than gasoline engines because *compression ratios* are higher, and because the absence of air throttling improves *volumetric efficiency*. Gasoline, gas (natural gas, propane), and diesel engines operate either on a **four-stroke cycle (Otto cycle)** or a **two-stroke cycle**. Most gasoline engines are of the four-stroke type, with operation as follows: 1) *intake*--piston moves down the cylinder, drawing in a fuel-air mixture through the intake valve; 2) *compression*--all valves closed, piston moves up, compressing the fuel-air mixture, and spark ignites mixture near top of stroke; 3) *power*--rapid expansion of hot combustion gases drives piston down, all valves remain closed; 4) *exhaust*--exhaust valve opens and piston returns, forcing out spent gases. The diesel four-stroke cycle differs in that only air is admitted on the intake stroke, fuel is injected at the top of the compression stroke, and the fuel-air mixture is ignited by the heat of compression rather than by an electric spark. The four-stroke-cycle engine has certain advantages over a two-stroke, which include: higher piston speeds, wider variation in speed and load, cooler pistons, no fuel lost through the exhaust, and lower fuel consumption. The two-stroke cycle eliminates the intake and exhaust strokes of the four-stroke cycle. As the piston ascends, it compresses the charge in the cylinder, while simultaneously drawing a new fuel-air charge into the crankcase, which is air-tight. (In the diesel two-stroke cycle, only air is drawn in; the fuel is injected at the top of the compression stroke). After ignition, the piston descends on the power stroke, simultaneously compressing the fresh charge in the crankcase. Toward the end of the power stroke, intake ports in the piston skirt admit a new fuel-air charge that sweeps exhaust products from the cylinder through exhaust ports; this means of flushing out exhaust gases is called "scavenging." Because the crankcase is needed to contain the intake charge, it cannot double as an oil reservoir. Therefore, lubrication is generally supplied by oil that is pre-mixed with the fuel. An important advantage of the two-stroke-cycle engine is that it offers twice as many power strokes per cycle and, thus, greater output for the same displacement and speed. Because two-stroke engines are light in relation to their output, they are frequently used

where small engines are desirable, as in chain saws, outboard motors, and lawn mowers. Many commercial, industrial, and railroad diesel engines are also of the two-stroke type. Gas turbines differ from conventional internal combustion engines in that a continuous stream of hot gases is directed at the blades of a rotor. A compressor section supplies air to a combustion chamber into which fuel is sprayed, maintaining continuous combustion. The resulting hot gases expand through the turbine unit, turning the rotor and driveshaft. See *turbine*.

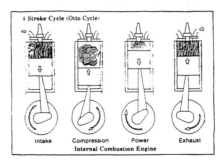

Internal Combustion Engine

International System of Units--see *SI*.

inverse emulsion--see *emulsion*.

ion--electrically charged atom, or group of atoms, that has lost or gained electrons. Electron loss makes the resulting particle positive, while electron gain makes the particle negative.

isomer--molecule having the same molecular formula as another molecule, but having a different structure and, therefore, different properties. As the carbon atoms in a molecule increase, the number of possible combinations, or isomers, increases sharply. For example, octane, an 8-carbon-atom molecule, has 18 isomers; decane, a 10-carbon-atom molecule, has 75 isomers.

normal butane

isobutane

isooctane--an *isomer* of octane (C_8H_{18}) having very good *antiknock* properties. With a designated octane number of 100, isooctane is used as a standard for determining the *octane number* of gasolines.

isoparaffin--branched *isomer* of a straight-chain *paraffin* molecule.

isoprene rubber--see *polyisoprene rubber*.

isothermal--pertaining to the conduct of a process or operation of equipment under conditions of constant temperature.

ISO viscosity classification system--international system, approved by the International Standards Organization (ISO), for classifying industrial lubricants according to *viscosity*. Each ISO viscosity grade number deisgnation corresponds to the mid-point of a viscosity range expressed in centistokes (cSt) at 40°C. For example, a lubricant with an ISO grade of 32 has a viscosity within the range of 28.8--35.2 cSt, the mid-point of which is 32.

jet fuel--see *turbo fuel*.

joule--unit of energy in the *Systeme International*, equal to the work done when the point of application of a force of one *newton* is displaced a distance of one meter in the direction of the force.

journal--that part of a shaft or axle which rotates in or against a *bearing*.

°K (Kelvin)--see *temperature scales*.

kauri-butanol (KB) value--a measure of the solvency of a *hydrocarbon*; the higher the kauri-butanol value, the greater the general solvent power of the hydrocarbon. Under test conditions prescribed in test method ASTM D 1133, a hydrocarbon sample is added to a standard solution of kauri gum in butyl alcohol (butanol) until sufficient kauri gum precipitates to blur vision of 10-point type viewed through the flask. When used in varnish, lacquer, and enamel formulations, a hydrocarbon *diluent* with a high kauri-butanol value dissolves relatively large quantities of solids.

Kelvin (°K)--see *temperature scales*.

kerosene--relatively colorless light *distillate*, heavier than gasoline (see *distillation*). It is used for lighting and heating, and as a fuel for some *internal combustion engines*.

kilocalorie--see *calorie*.

kilowatt-hour--unit of work or energy, equivalent to the energy expended in one hour at a steady rate of one kilowatt. Total kilowatt-hours (kwh) consumed by one 100-watt light bulb burning for 150 hours can be calculated as follows: 100 watts x 150 hours = 15,000 watt-hours = 15 kwh.

kinematic viscosity--*absolute viscosity* of a fluid divided by its density at the same temperature of measurement. It is the measure of a fluid's resistance to flow under gravity, as determined by test method ASTM D 445. To determine kinematic viscosity, a fixed volume of the test fluid is allowed to flow through a calibrated capillary tube (*viscometer*) that is held at a closely controlled temperature. The kinematic viscosity, in centistokes (cSt), is the product of the measured flow time in seconds and the calibration constant of the viscometer. See *viscosity*.

knock--in the cylinder of a spark-ignited *internal combustion engine*, premature explosion of a portion of the air-fuel mixture, independent of spark plug ignition, as a result of excessive heat buildup during compression. The high local pressures resulting from the explosion are the source of the objectionable clatter or ping associated with knock. Knock reduces efficiency and can be destructive to engine parts. High-octane gasolines resist knocking. Also called **detonation**. See *octane number, pre-ignition*.

laminating strength--bonding strength of a petroleum wax used as an adhesive between layers of paper or foil, measured in terms of the specific number of grams of force required to peel the layers apart. It is expressed in grams per inch of width of the layers. *Sealing strength*, similar in meaning to laminating strength, may also be measured in this manner.

lard oil--*fatty oil* used for compounding. See *compounded oil*.

latent heat--quantity of heat absorbed or released by a substance undergoing a change of state (e.g., ice changing to liquid water, or water to steam) without change of temperature. At standard atmospheric pressure, the latent heat of vaporization of water is 2256 kJ/kg (970 Btu/lb); this is the amount of energy required to convert water at 100°C (212°F) to steam at the same temperature. Conversely, when steam condenses at 100°C, the same amount of heat is released.

launching lubricant--lubricant applied to inclined launching guides, or ways, to facilitate launching of a ship. Two separate lubricants are usually used: a firm, abrasion-resistant base coat, plus a softer, low-friction slip coat.

LC$_{50}$--lethal concentration, 50% mortality; a measure of inhalation toxicity. It is the concentration in air of a volatile chemical compound at which half the test population of an animal species dies when exposed to the compound. It is expressed as parts per million by volume of the toxicant per million parts of air for a given exposure period.

LCN--light crude naphtha. See *naphtha*.

LD$_{50}$--lethal dose, 50% mortality, a general measurement of toxicity. It is the dose of a chemical compound that, when administered to laboratory animals, causes death in one-half the test population. It is expressed in milligrams of toxicant per kilogram of animal weight. The route of administration, may be oral, epidermal, or intraperitoneal.

lead alkyl--any of several lead compounds used to improve *octane number* in a gasoline. The best known is **tetraethyl lead** (TEL), Pb $(C_2H_5)_4$. Another is **tetramethyl lead** (TML), Pb $(CH_3)_4$. Other compounds have varying proportions of methyl *radicals* (CH_3) and ethyl radicals (C_2H_5). Use of lead compounds in motor gasoline is being phased out for environmental reasons. Beginning with the 1980-model year, all new U.S. and foreign-made cars sold in the U.S. required *unleaded gasoline*.

lead naphthenate, lead oleate--lead soaps that serve as mild *EP additives*. These additives are seldom used in modern lubricants because of environmental considerations.

lead scavenger--see *scavenger*.

lean and rich octane number--expression of the *antiknock* value of an *aviation gasoline* at lean air-fuel mixtures (relatively low concentration of fuel) and rich air-fuel mixtures, respectively. A grade designation of 80/87 means that at lean mixtures the fuel performs like an 80-octane gasoline and at rich mixtures, like an 87-octane gasoline. See *performance number*.

light crude naphtha--see *naphtha*.

light ends--low-boiling-point *hydrocarbons* in gasoline having up to five carbon atoms, e.g., butanes, butenes, *pentanes*, pentenes, etc. Also, any extraneous low-boiling fraction in a refinery process stream.

limestone--porous, sedimentary rick composed chiefly of calcium carbonate; sometimes serves as a *reservoir* rock for petroleum.

linear paraffin--see *normal paraffin*.

liquefied natural gas--see *LNG*.

liquefied petroleum gas--see *LPG*.

lithium soap grease--see *grease*.

LNG (liquefied natural gas)-- natural gas that has been liquefied at extremely low temperature. It is stored or transported in insulated tanks capable of sustaining the high pressure developed by the product at normal ambient temperatures.

load wear index (LWI)--measure of the relative ability of a lubricant to prevent wear under applied loads; it is calculated from data obtained from the Four-Ball EP Method. Formerly called **mean Hertz load**. See *four-ball method*.

local effect--toxic effect that is limited to the area of the body (commonly the skin and eyes) that has come into contact with a toxicant.

Lovibond tintometer--device for measuring the color of a petroleum product, particularly *petrolatums*. The melted petrolatum is contained in a cell and the color is compared with a series of yellow and red Lovibond glasses. The length of the cell and the color standards that give the best match are reported. See *color scale*.

lower flammable limit--the concentration of a flammable vapor mixed with air that will just propagate flame, that is, continue to burn. See *explosive limits*.

low-temperature corrosion--see *cold-end corrosion*.

LP gas--see *LPG*.

LPG (liquefied petroleum gas)-- *propane* or (less commonly) *butane*, obtained by extraction from *natural gas* or from refinery processes. LPG has a *vapor pressure* sufficiently low to permit compression and storage in a liquid state at moderate pressures and normal ambient temperatures. Pressurized in metal bottles or tanks. LPG is easily handled and readily lends itself to a variety of applications as a fuel, refrigerant, and propellant in packaged aerosols. LPG is also called **LP gas** and *bottled gas*. See *natural gas liquids*.

lubrication--control of friction and wear by the introduction of a friction-reducing film between moving surfaces in contact. The lubricant used may be a fluid, solid, or plastic substance. For principles of lubrication, see *boundary lubrication, full-fluid-film lubrication, ZN/P curve*. For methods of lubrication, see *centralized lubrication, circulating lubrication, oil mist lubrication, once-through lubrication*.

lubricity--ability of an oil or grease to lubricate; also, called **film strength**. Lubricity can be enhanced by *additive* treatment. See *compounded oil*.

luminometer number--measure of the flame radiation characteristics of a turbine fuel, as determined by test method ASTM D 1740. A sample of fuel is burned in a luminometer lamp, and the temperature rise, at a specified flame radiation value, is compared with the corresponding temperature rise of reference fuels. The higher the luminometer number, the lower the flame radiation and the better the combustion characteristics.

LWI--see *load wear index*.

machine oil--see *once-through lubrication*.

marquenching--see *quenching*.

martempering--see *quenching*.

mass spectrometer--apparatus for rapid quantitative and qualitative analysis of hydrocarbon compounds in a petroleum sample. It utilizes the principle of accelerating molecules in a circular path in an electrical field. The compounds are separated by centrifugal force, with the molecules having a greater mass (weight) being thrown to the outer periphery of the path. Quantitative measurements are accomplished by use of either a photographic plate or electronic determination of the relative proportions of each type of particle of a given mass.

mastic--any of various semi-solid substances, usually formulated with rubber, other *polymers*, or oxidized *asphalt*; commonly used as a tile adhesive caulking, and a sound-reducing treatment on various surfaces.

MC asphalt--see *cutback asphalt*.

mean Hertz load--see *load wear index*.

mechanical stability--see *structural stability*.

melting point of wax--temperature at which a sample of wax either melts or solidifies from the solid or liquid state, respectively, depending on the *ASTM* test used. Low melting point generally indicates low *viscosity*, low *blocking point*, and relative softness.

mercaptan--any of a generic series of malodorous, toxic sulfur compounds occurring in crude oil. Mercaptans are removed from most petroleum products by refining, but may be added to natural gas and *LPG* in very low concentrations to give a distinctive warning odor.

merit rating--arbitrary graduated numerical rating commonly used in evaluating engine deposit levels when testing the *detergent-dispersant* characteristics of an engine oil. On a scale of 10-0, the lower the number, the heavier the deposits. A less common method of evaluating engine cleanliness is *demerit rating*. See *engine deposits*.

metal wetting--see *polar compound*.

methane--a light, odorless, flammable gas (CH_4); the chief constituent of natural gas.

methanol--the lowest molecular weight *alcohol* (CH_3OH). Also called **methyl alcohol** and **wood alcohol**.

methyl alcohol--see *methanol*.

metric system--international decimal system of weights and measures based on the meter and kilogram. The following table presents common metric units and their U.S. equivalents:

Metric Unit		Equivalent in U.S. Customary Unit
1 meter (m)	=	3.28 feet
1 centimeter (cm)	=	0.394 inches
1 kilometer (km)	=	0.621 miles
1 kilogram (kg)	=	2.205 pounds
1 gram (g)	=	0.353 ounces
1 liter (L)	=	1.056 quarts (liquid)

microcrystalline wax, microwax--see *wax (petroleum)*.

mid-boiling point--see *distillation test*.

middle distillate--see *distillate*.

military specifications for engine oils--There are six military specifications for engine oils. Some of these specifications are obsolete, but are still commonly used to designate required engine oil performance levels. To qualify under a military specification, an oil must meet minimum requirements in laboratory engine tests. In most cases these are the same *ASTM* tests used to define *API Engine Service Categories*. On the next page is a listing of the military specifications and API Service equivalents for engine oils in order of ascending quality (the API Service equivalents of obsolete military specifications remain valid).

mineral oil--any petroleum oil, as contrasted to animal or vegetable oils. Also, a highly refined petroleum *distillate*, or *white oil*, used medicinally as a laxative.

Mil. Spec.	Description
MIL-L-2104A (API CA)	obsolete, superseded by MIL-L-2104B; describes an oil type still used by some fleet owners for older. less critical equipment
Supplement I (API CB)	obsolete; describes oil type similar to MIL-L-2104A, but tested under more severe conditions
MIL-L-2104B (API CC)	obsolete, superseded by MIL-L-46152, but still widely used as a standard
MIL-L-46152 (API SE-CC. SF-CC)	supersedes MIL-L-2104B
MIL-L-45199B (API CD)	obsolete, superseded by MIL-L-2104C; the specification is essentially the same as the *Series 3* specification established by Caterpillar Tractor Company
MIL-L-2104C (API CD)	supersedes MIL-L-45199B; covers oil requirements for heavy trucks, tanks, and other tactical military vehicles. and includes gasoline engine performance level between API SC and SD

mineral seal oil--*distillation* fraction between *kerosene* and *gas oil,* widely used as a solvent oil in gas absorption processes (see *absorber oil*), as a lubricant for the rolling of metal foil, and as a base oil in many specialty formulations. Mineral seal oil takes its name--not from any sealing function--but from the fact that it originally replaced oil derived from seal blubber for use as an illuminant for signal lamps and lighthouses.

mineral spirits--*naphthas* of mixed hydrocarbon composition and intermediate volatility, within the boiling range of 149°C (300°F) to 204°C (400°F) and with a *flash point* greater than 38°C (100°F); widely used as solvents or thinners in the manufacture of cleaning products, paints, lacquers, inks, and rubber. Also used uncompounded for cleaning metal and fabrics.

miscible--capable of being mixed in any concentration without separation of phases; e.g., water and ethyl alcohol are miscible. See *immiscible.*

mist lubrication-- see *oil mist lubrication.*

mobilometer--device for measuring the relative consistency or resistance to flow of fluid grades of grease too soft to be tested in the penetrometer. See *consistency (grease).*

mold lubricant--a compound, often of petroleum origin, for coating the interiors of molds for glass and ceramic products. The mold lubricant facilitates removal of the molded object from the mold, protects the surface of the mold, and reduces or eliminates the need for cleaning it. Also called **release agent.**

moly, molysulfide--see *molybdenum disulfide.*

molybdenum disulfide--a black, lustrous powder (MoS_2) that serves as a dry-film lubricant in certain high-temperature and high-vacuum applications. It is also used in the form of pastes to prevent *scoring* when assembling press-fit parts, and as an additive to impart residual lubrication properties to oils and greases. Molybdenum disulfide is often called **moly** or **molysulfide.**

monomer--see *polymer*.

Mooney scorch value--see *Mooney viscosity*.

Mooney viscosity--measure of the resistance of raw or unvulcanized rubber to deformation, as measured in a Mooney viscometer. A steel disc is embedded in a heated rubber specimen and slowly rotated. The resistance to the shearing action of the disc is measured and expressed as a Mooney viscosity value. Viscosity increases with continued rotation, and the time required to produce a specified rise in Mooney viscosity is known as the **Mooney scorch value**, which is an indication of the tendency of a rubber mixture to cure, or vulcanize, prematurely during processing.

Motor Octane Number--see *octane number*.

motor oil--see *engine oil*.

MS asphalt--see *emulsified anionic asphalt*.

mud--liquid circulated through the *borehole* during *rotary drilling*; it is used to bring cuttings to the surface, cool and lubricate the drill stem, protect against blowouts by holding back subsurface pressures, and prevent fluid loss by plastering the borehole wall. Mud formulations originally were suspensions of clay or other earth solid in water,

but today are more complex mixtures of liquids (not necessarily water), reactive solids, and inert solids.

multi-grade oil--engine oil that meets the requirements of more than one *SAE* (Society of Automotive Engineers) viscosity grade classification (see *SAE viscosity grades*), and may therefore be suitable for use over a wider temperature range than a single-grade oil. Multi-grade oils have two viscosity grade numbers indicating their lowest and highest classification, e.g., SAE 10W-40. The lower grade number indicates the relative fluidity of the oil in cold weather for easy starting and immediate oil flow. The higher grade number indicates the relative viscosity of the oil at high operating temperatures for adequate wear protection. The "W" means "winter" grade. Multi-grade oils generally contain *viscosity index (V.I.) improvers* that reduce the tendency of an oil to lose viscosity, or thin out, at high temperatures.

multi-purpose grease--high-quality grease that can be used in a variety of applications. See *grease*.

naphtha--generic, loosely defined term covering a range of light petroleum *distillates* (see *distillation*). Included in the naphtha classification are: *gasoline blending stocks*, *mineral spirits*, and a broad selection of petroleum *solvents*. In refining, the term **light crude naphtha** (LCN) usually refers to the first liquid distillation fraction, boiling range 32° to 100°C (90° to 175°F), while **heavy crude naphtha** is usually the second distillation fraction, boiling range 163° to 218°C (325° to 425°F).

naphthene--hydrocarbon characterized by saturated carbon atoms in a ring structure, and having the general formula CnH_2n; also called **cycloparaffin** or **cycloalkane**. Naphthenic lubricating oils have low *pour points*, owing to their very low wax content, and good solvency properties. See *hydrocarbon, saturated hydrocarbon*.

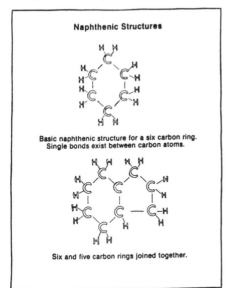

Naphthenic Structures

Basic naphthenic structure for a six carbon ring. Single bonds exist between carbon atoms.

Six and five carbon rings joined together.

naphthenic--see *naphthene*.

narrow cut--see *distillation test*.

National Formulary--see *NF*.

natural gas--naturally occurring mixture of gaseous *saturated hydrocarbons*, consisting of 80-95% methane (CH_4), lesser amounts of *propane, ethane*, and *butane*, and small quantities of nonhydrocarbon gases (e.g., nitrogen, helium). Natural gas is found in sandstone, limestone, and other porous rocks beneath the earth's surface, often in association with crude oil. Because of its high heating value and clean-burning characteristics, natural gas is widely used as a fuel. The heavier hydrocarbons in natural gas can be extracted, through compression or absorption processes, to yield *LPG* (propane or butane), *natural gasoline*, and raw materials for *petrochemical* manufacture. See *natural gas liquids*.

natural gas liquids--hydrocarbons extracted from *natural gas*: primarily *LPG* (propane or butane) and *natural gasoline*, the latter being commonly blended with crude-derived gasoline to improve volatility. Natural gas liquids can be separated from the lighter hydrocarbons of natural gas by compression (the gas is compressed and cooled until the heavier hydrocarbons liquefy) or by absorption (the gas is mixed with a petroleum distillate, such as kerosene, which absorbs, or dissolves, the heavier hydrocarbons).

natural gasoline--liquid hydrocarbons recovered from wet natural gas; also called casinghead gasoline. See *natural gas liquids*.

natural rubber--resilient *elastomer* generally prepared from the milky sap, or latex, of the rubber tree (*hevea brasilensis*). Natural rubber possesses a degree of tack (adhesive properties) not inherent in most *synthetic rubbers*. It may be used unblended in large tires for construction and agricultural equipment and airplanes, where low rolling resistance and low heat buildup are of greater importance than wear resistance. In the manufacture of tires for highway vehicles, natural rubber may be added to synthetic rubber to provide the necessary tack.

NBR--see *nitrile rubber*.

neoprene rubber (CR)--*synthetic rubber*, a chloroprene *polymer*, with excellent resistance to weather, oil, chemicals, and flame. Widely used for electrical cable insulation, industrial hose, adhesives, shoe soles, and paints.

neutralization number--also called neut number, an indication of the acidity or alkalinity of an oil; the number is the weight in milligrams of the amount of acid (hydrochloric acid [HCl] or base (po-

tassium hydroxide [KOH]) required to neutralize one gram of the oil, in accordance with test method ASTM D 664 (potentiometric method) or ASTM D 974 (colorimetric method). Strong acid number is the weight in milligrams of base required to titrate a one-gram sample up to a *pH* of 4; total acid number is the weight in milligrams of base required to neutralize all acidic constituents. Strong base number is the quantity of acid, expressed in terms of the equivalent number of milligrams of KOH, required to titrate a one-gram sample to a pH of 11; total base number is the milligrams of acid, expressed in equivalent milligrams of KOH, to neutralize all basic constituents.   If the neutralization number indicates increased acidity (i.e., high acid number) of a used oil, this may indicate that oil oxidation, additive depletion, or a change in the oil's operating environment has occurred.

newton--in the *Systeme International*, the unit of force required to accelerate a mass of one kilogram one meter per second.

Newtonian fluid--fluid, such as a *straight mineral oil*, whose *viscosity* does not change with rate of flow. See *shear stress*.

NF (National Formulary)--listing of drugs, drug formulas, quality standards, and tests published by the United States Pharmacopeial

Convention, Inc., which also publishes the *USP* (United States Pharmacopeia). The purpose of the NF is to ensure the uniformity of drug products and to maintain and upgrade standards of drug quality, packaging, labeling, and storage. In 1980, all NF responsibility for *white oil* classification was transferred to the USP.

Niemann four-square gear oil test--see *FZG four-square gear oil test*.

nitration grade--term for *toulene*, *xylene*, or *benzene* refined under close controls for very narrow boiling range and high purity. Nitration-grade specifications are given in test methods ASTM D 841, D 843, and D 835, respectively.

nitrile rubber (NBR)--*synthetic rubber* made by the copolymerization of butadiene and acrylonitrile. It resists heat, oil, and fuels; hence, is used in gasoline and oil hose, and in tank linings. Originally called **Buna-N**. See *polymer*.

nitrogen blanket--see *gas blanket*.

nitrogen oxides (NOx)--emissions, from man-made and natural sources, of nitric oxide (NO), with minor amounts of nitrogen dioxide (NO_2). NOx are formed whenever fuel is burned at high temperatures in air, from nitrogen in the air as well as in the fuel. Motor vehicles

and stationary combustion sources (furnaces and boilers) are the primary man-made sources, although automotive emission controls are reducing the automobile's contribution. Natural emissions of NOx arise from bacterial action in the soil. NOx can react with hydrocarbons to produce smog. See *catalytic converter, emissions (automotive), emissions (stationary source), pollutants, hydrocarbon emissions*.

NLGI (National Lubricating Grease Institute)--trade association whose main interest is grease and grease technology. NLGI is best known for its system of rating greases by penetration. See *NLGI consistency grades, penetration (grease)*.

NLGI consistency grades--simplified system established by the National Lubricating Grease Institute (NLGI), for rating the consistency of grease. See *penetration (grease)*. The following are the NLGI consistency grades:

NLGI Consistency Grades	
NLGI Grade	ASTM Worked Penetration, mm/10
000	445-475
00	400-430
0	355-385
1	310-340
2	265-295
3	220-250
4	175-205
5	130-160
6	85-115

non-Newtonian fluid--fluid, such as a grease or a *polymer*-containing oil (e.g., multi-grade oil), in which *shear stress* is not proportional to *shear rate*. See *Brookfield viscosity*.

non-soap thickener--see *grease*.

non-volatiles--see *solids content*.

normal paraffin--hydrocarbon consisting of unbranched molecules in which any carbon atom is attached to no more than two other carbon atoms; also called **straight chain paraffin** and **linear paraffin**. See *isoparaffin, paraffin*.

occupational exposure limit (OEL)--the *time-weighted average* concentration of a material in air for an eight-hour workday, 40-hour workweek to which nearly all workers may be exposed repeatedly without adverse effect. Also called **threshold limit value (TLV)**.

Occupational Safety and Health Act of 1970--the main legislation affecting health and safety in the workplace. It created the Occupational Safety and Health Administration (OSHA) in the Department of Labor, and the National Institute for Occupational Safety and Health in the Department of Health and Human Services (formerly Department of Health, Education, and Welfare).

OCS--see *outer continental shelf*.

octane number--expression of the *antiknock* properties of a gasoline,

relative to that of a standard reference fuel. There are two distinct types of octane number measured in the laboratory: **Research Octane Number (RON)** and **Motor Octane Number (MON)**, determined in accordance with ASTM D 2699 and D 2700, respectively. Both the RON and MON tests are conducted in the same laboratory engine, but RON is determined under less severe conditions, and is therefore numerically greater than MON for the same fuel. The average of the two numbers--(RON + MON)/2--is commonly used as the indicator of a gasoline's road antiknock performance. The gasoline being tested is run in a special single-cylinder engine, whose compression ratio can be varied (the higher the compression ratio, the higher the octane requirement). The knock intensity of the test fuel, as measured by a knockmeter, is compared with the knock intensities of blends of *isooctane* (as signed a knock rating of 100) and *heptane* (with a knock rating of zero), measured under the same conditions as the test fuel. The percentage, by volume, of the isooctane in the blend that matches the characteristics of the test fuel is designated as the octane number of the fuel. For example, if the matching blend contained 90% isooctane, the octane number of the test fuel would be 90. In addition to the laboratory tests for RON and MON, there is a third method. **Road Octane Number**, which is conducted in a specially equipped test car by individuals trained to hear trace levels of engine knock. See *antiknock compounds, knock.*

odorless solvents--*solvents*, generally *mineral spirits*, that are synthesized by *alkylation* and refined to remove odorous *aromatics* and sulfur compounds; there remains, however, a relatively low level of odor inherent in the hydrocarbons. Odorless solvent applications include dry cleaning and odorless paint manufacture.

odor panel--a group of individuals trained to identify and rate odors, in order to check the odor quality of solvents, waxes, etc.

OEL--see *occupational exposure limit.*

oil content of petroleum wax--a measure of wax refinement, under conditions prescribed by test method ASTM D 721. The sample is dissolved in methyl ethyl ketone, and cooled to--32°C (-26°F) to precipitate the wax, which is then filtered out. The oil content of the remaining filtrate is determined by evaporating the solvent and weighing the residue. Waxes with an oil content generally of 1.0 mass percent or less are known as *refined waxes.* Refined waxes are harder and have greater resistance to blocking (see *blocking point*) and staining than waxes with higher oil content. Waxes with an oil content up to 3.0 mass percent are general-

ly referred to as *scale waxes*, and are used in applications where the slight color, odor, and taste imparted by the higher oil content can be tolerated. Semi-refined *slack waxes* may have oil contents up to 30 mass percent, and are used in non-critical applications. The distinction between scale and slack waxes at intermediate oil content levels (2-4 mass percent) is not clearly defined, and their suitability for particular applications depends upon properties other than oil content alone. See *wax.*

oiliness agent--*polar compound* used to increase the *lubricity* of a lubricating oil and aid in preventing wear and *scoring* under conditions of *boundary lubrication.*

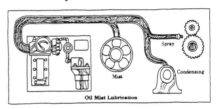

Oil Mist Lubrication

oil mist lubrication--type of *centralized lubrication* that employs compressed air to transform liquid oil into a mist that is then distributed at low pressure to multiple points of application. The oil mist is formed in a "generator", where compressed air is passed across an orifice, creating a pressure reduction that causes oil to be drawn from a reservoir into the airstream. The resulting mist (composed of fine droplets on the average of 1.5 microns) is distributed through feed lines to various application points. Here, it is reclassified, or condensed, to a liquid, spray, or coarser mist by specialized fittings, depending on the lubrication requirements. Oils for use in a mist lubrication system are formulated with carefully selected *base stocks* and *additives* for maximum delivery of oil to the lubrication points and minimal coalescence of oil in the feed lines.

oil shale--shale containing a rubbery hydrocarbon known as kerogen. When shale is heated, the kerogen vaporizes and condenses as a tar-like oil called **shale oil**, which can be upgraded and refined into products in much the same way as liquid petroleum. There are large oil shale deposits in the U.S., the richest being in Colorado, Utah, and Wyoming.

olefin--any of a series of unsaturated, relatively unstable hydrocarbons characterized by the presence of a double bond between two carbon atoms in its structure, which is commonly straight-chain or branched. The double bond is chemically active and provides a focal point for the addition of other reactive elements, such as oxygen. Due to their ease of oxidation, olefins are undesirable in petroleum solvents and lube oils. Examples of olefins are: *ethylene* and *propylene*. See *hydrocarbon, unsaturated hydrocarbon*.

olefin oligomer--*synthetic lubricant* base, formed by the polymerization of *olefin* monomers (see *polymer*); properties include good *oxidation stability* at high temperatures, good *hydrolytic stability*, good compatibility with *mineral oils*, and low *volatility*. Used in turbines, compressors, gears, automotive engines, and electrical applications.

once-through lubrication--system of lubrication in which the lubricant is supplied to the lubricated part at a minimal rate and is not returned or recirculated. Lubrication by oil can, mechanical lubricator, centralized grease system, lubricating device, oil mist, etc., is done on a once-through basis. Since the lubricant is not recovered, high *oxidation stability* and long service life are usually not necessary, but *viscosity* and other properties may be very important. Oils that meet the moderate requirements of once-through lubrication are known variously as **machine oils** and **general-purpose oils**. See *centralized lubrication, oil mist lubrication*.

OPEC (Organization of Petroleum Exporting Countries)--group of oil-producing nations founded in 1960 to advance member interests in dealings with industrialized oil-consuming nations. The 13 OPEC members are: Algeria, Ecuador, Gabon, Indonesia, Iran, Iraq, Kuwait, Libya, Nigeria, Qatar, Saudi

Arabia, United Arab Emirates, and Venezuela. Rising world oil demand, tight world oil supplies, and declining U.S. oil and gas production have enabled OPEC to dramatically increase the price of its oil exports since 1973.

open cup--see *Cleveland open cup, Tag open cup.*

orchard spray oil--petroleum oil suitable for emulsifying with water to form an insecticide spray that kills orchard pests by suffocation. When applied to fruit trees as directed, it has proved highly effective in the control of certain insects that attack citrus, apples, pears, peaches, nuts, and other orchard crops. The phytotoxicity (harmfulness to plants) depends on the boiling range and purity of the oil. Purity is broadly defined by the *unsulfonated residue* of the oil. Oils with an unsulfonated residue of 92% or higher can be used in sensitive applications, such as verdant, or summer sprays when trees are in leaf. These are known as "superior" spray oils. Oils with lower unsulfonated residues--at least 80%--are called "regular" spray oils, and are limited to application only in the dormant phase of plant growth.

organic compound--chemical substance containing carbon and hydrogen; other elements, such as nitrogen oxygen, may also be present. See *hydrocarbon, inor-*

ganic compound.

Organization of Petroleum Exporting Countries--see *OPEC.*

organosol--see *plastisols and organosols.*

OSHA (Occupational Safety and Health Administration)--see *Occupational Safety and Health Act of 1970.*

Otto cycle--four-stroke engine cycle. See *internal combustion engine.*

outer continental shelf--the part of the continental margin that slopes gradually away from the shore to a point where a much steeper drop begins. The outer continental shelf may extend from a few miles to several hundred miles from the shore. It is much wider off the U.S. East and Gulf coasts, for example, than off the West Coast. Much of the remaining U.S. oil and gas resources are believed to lie beneath the outer continental shelf.

overhead--the distillation fraction removed as vapor or liquid from the top of a distillation column, e.g., a pipe still. See *distillation.*

oxidation--the chemical combination of a substance with oxygen. All petroleum products are subject to oxidation, with resultant degradation of their composition and performance. The process is accel-

erated by heat, light, metal catalysts (e.g., copper), and the presence of water, acids, or solid contaminants. The first reaction products of oxidation are organic peroxides. Continued oxidation, cata-
lyzed by peroxides, forms alcohols,
aldehydes, ketones, and organic
acids, which can be further oxidized to form high-molecular-
weight, oil-insoluble *polymers*;
these settle out as sludges, varnishes, and gums that can impair equipment operation. Also, the organic
acids formed from oxidation are
corrosive to metals. Oxidation
resistance of a product can be
improved by careful selection of
base stocks (*paraffins* have greater
oxidation resistance than
naphthenes), special refining methods, and addition of *oxidation
inhibitors*. Also, oxidation can be
minimized by good maintenance of
oil and equipment to prevent contamination and excessive hat. See
oxidation stability.

oxidation inhibitor--substance
added in small quantities to a petroleum product to increase its oxidation resistance, thereby lengthening
its service or storage life; also
called **anti-oxidant**. An oxidation
inhibitor may work in one of three
ways: 1) by combining with and
modifying peroxides (initial oxidation products) to render them harm-
less, 2) by decomposing the peroxides, or 3) by rendering an oxidation catalyst (metal or metal ions)
inert. See *oxidation*.

oxidation stability--resistance of a
petroleum product to oxidation;
hence, a measure of its potential
service or storage life. There are a
number of ASTM tests to determine the oxidation stability of a
lubricant or fuel, all of which are
intended to simulate service conditions on an accelerated basis. In
general, the test sample is exposed
to oxygen or air at an elevated
temperature, and sometimes to
water or catalysts (usually iron or
copper). Depending on the test,
results are expressed in terms of
the time required to produce a
specified effect (such as a pressure
drop), the amount of sludge or gum
produced, or the amount of oxygen
consumed during a specified period.

oxidized asphalt--also called **blown
asphalt**. See *asphalt*.

pale oil--straight naphthenic mineral oil, straw, or pale yellow in
color, used as a once-through
lubricant and in the formulation of
process oils. See *naphthene,
straight mineral oil, once-through
lubrication*.

paraffin--hydrocarbon identified by
saturated straight (normal) or
branched (iso) carbon chains. The
generalized paraffinic molecule can
be symbolized by the formula
CnH_2n+_2. Paraffins are relatively
non-reactive and have excellent
oxidation stability. In contrast to
naphthenic (see *naphthene*) oils,

paraffinic lube oils have relatively high wax content and *pour point*, and generally have a high *viscosity index (V.I.)*. Paraffinic solvents are generally lower in solvency than naphthenic or *aromatic* solvents. See *hydrocarbon, normal paraffin, isoparaffin, saturated hydrocarbon*.

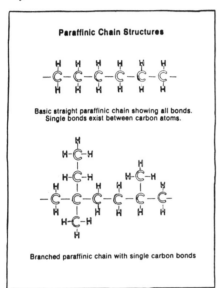

Paraffinic Chain Structures

Basic straight paraffinic chain showing all bonds. Single bonds exist between carbon atoms.

Branched paraffinic chain with single carbon bonds

paraffinic--see *paraffin*.

paraffin wax--petroleum-derived wax usually consisting of high-molecular-weight *normal paraffins*; distinct from other natural waxes, such as beeswax and carnauba wax (palm tree), which are composed of high-molecular-weight *esters*, in combination with high-molecular-weight acids, alcohols, and hydrocarbons. See *wax (petroleum)*.

partial pressure--pressure exerted by a single component of a gaseous mixture. The sum of the partial pressures in a gaseous mixture equals the total pressure. The partial pressure of a substance is a function both of its *volatility*, or *vapor pressure*, and its concentration.

particulates--atmospheric particles made up of a wide range of natural materials (e.g., pollen, dust, resins), combined with man-made pollutants (e.g., smoke particles, metallic ash); in sufficient concentrations, particulates can be a respiratory irritant. Primary sources of man-made particulate emissions are industrial process losses (e.g., from cement plants) and stationary combustion sources. Motor vehicles contribute a relatively minor amount of particulates. See *emissions (stationary source), pollutants*.

pascal (Pa)--in the *Systeme International*, a unit of pressure equivalent to a force of one *newton (n)* applied to an area of one square meter.

paving asphalt--see *asphalt cement*.

PCB--polychlorinated biphenyl, a class of synthetic chemicals consisting of an homologous series of compounds beginning with monochlorobiphenyl and ending with decachlorobiphenyl. PCB's do not occur naturally in petroleum, but have been found as contaminants in used oil. PCB's have been legally designated as a health hazard, and any oil so contaminated must be handled in strict accordance with state and federal regulations.

PCV--see *positive crankcase ventilation*.

PE--see *polyethylene*.

penetration (asphalt)--method for determining the penetration, or consistency, of semi-solid and solid bituminous (see *bitumen*) materials, in a *penetrometer* under conditions prescribed by test method ASTM D 5. The test sample is melted and cooled under controlled conditions, then a weighted standard needle is positioned at the surface of the sample and allowed to penetrate it, by means of gravity, for a specified time. The penetration is measured in tenths of a millimeter. See *viscosity (asphalt)*.

penetration (grease)--measure of the consistency of a grease, utilizing a *penetrometer*. Penetration is reported as the tenths of a millimeter (penetration number) that a standard cone, acting under the influence of gravity, will penetrate the grease sample under test conditions prescribed by test method ASTM D 217. Standard test temperature is 25°C (77°F). The higher the penetration number, the softer the grease. **Undisturbed penetration** is the penetration of a

grease sample as originally received in its container. **Unworked penetration** is the penetration of a grease sample that has received only minimal handling in transfer from its original container to the test apparatus. **Worked penetration** is the penetration of a sample immediately after it has been subjected to 60 double strokes in a standard grease worker; other penetration measurements may utilize more than 60 strokes. **Block penetration** is the penetration of *block grease* (grease sufficiently hard to hold its shape without a container).

penetration (wax)--measure of the consistency of a petroleum wax, utilizing a *penetrometer*. Penetration is reported as the depth, in tenths of a millimeter, to which a standard needle penetrates the wax under conditions described in test method ASTM D 1321. Prior to penetration, the wax sample is heated to 17°C (30°F) above its *congealing point*, air cooled, then conditioned at test temperature in a water bath, where the sample remains during the penetration test. The test temperature may range from 25° to 75°C (77° to 130°F).

penetration grading (asphalt)--classification system for *asphalt cement*, defined in AASHTO (American Association of State Highway Transportation Officials) Specification M20, and based on tests for *penetration (asphalt)*, *flash point*, ductility, purity, etc., as specified in test method ASTM D 946. There are five standard grades, ranging from hard to soft: 40-50, 60-70, 85-100, 120-150, and 200-300. Asphalt cement is also classified by *viscosity grading*.

penetrometer--apparatus for measuring the consistency of lubricating grease or asphalt. A standard cone (for grease) or needle (for wax or asphalt) is lowered onto a test sample, under prescribed conditions, and the depth of penetration is measured. See *mobilometer, penetration (asphalt), penetration (grease), penetration (wax)*.

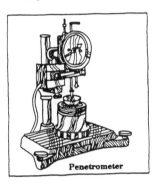

Penetrometer

Pensky-Martens closed tester--apparatus used in determining the *flash point* of *fuel oils* and *cutback asphalt*, under conditions prescribed by test method ASTM D 93. The test sample is slowly heated in a closed cup, at a specified constant rate, with continual stirring. A small flame is introduced into the cup at specified intervals through shuttered openings. The lowest temperature at which the vapors above the sample briefly ignite is the flash point. See *Tag closed tester*.

pentane--saturated paraffinic hydrocarbon (C_5H_{12}); it is a colorless, volatile liquid, normally blended into gasoline. See *paraffin, saturated hydrocarbon*.

pentane insolubles--see *insolubles*.

performance number--expression of the *antiknock* properties of an *aviation gasoline* with a Motor Octane Number higher than 100. The laboratory procedure for determining performance number is the same as that for Motor Octane Number (see *octane number*), except that performance number is based on the number of milliliters of tetraethyl lead (see *lead alkyl*) in the reference *isooctane* blend that matches the antiknock characteristics of the test fuel. A 100/130 grade aviation gasoline has an octane number of 100 at lean fuel mix and a performance number of 130 at rich mix. See *lean and rich octane numbers*.

peroxide--any compound containing two linked oxygen atoms (e.g., Na_2O_2) that yields hydrogen peroxide (H_2O_2) when reacted with acid; also, H_2O_2 itself. Relatively unstable, peroxides are strong oxidizing agents and, when present in lubricating oils, can accelerate oil oxidation and promote bearing corrosion. See *oxidation*.

pesticide--any chemical substance intended to kill or control pests. Common pesticides are : insecticides, rodenticides, herbicides, fungicides, and bactericides. Petroleum products and their petrochemical derivatives are important in the formulation of many types of pesticides. Specialized petroleum oils are used to kill insects by suffocation (see *orchard spray oil*); other petroleum products serve as solvents or *diluents* for the active component.

petrochemical--any chemical derived from crude oil, crude products, or natural gas. A petrochemical is basically a compound of carbon and hydrogen, but may incorporate many other elements. Petrochemicals are used in the manufacture of numerous products such as *synthetic rubber*, synthetic fibers (such as nylon and polyester), plastics, fertilizers, paints, detergents, and pesticides.

petrolatum--semi-solid, noncrystalline hydrocarbon, pale to yellow in color, composed primarily of high-molecular-weight waxes; used in lubricants, rust preventives, and medicinal ointments. See *wax (petroleum)*.

petroleum--term applied to crude oil and its products; also called

rock oil. See *crude oil*.

pH--measure of the acidity or alkalinity of an aqueous solution. The pH scale ranges from 0 (very acidic) to 14 (very alkaline), with a pH of 7 indicating a neutral solution equivalent to the pH of distilled water. See *neutralization number*.

phenol--white, crystalline compound (C_6H_5OH) derived from *benzene*; used in the manufacture of phenolic *resins*, weed killers, plastics, disinfectants; also used in *solvent extraction*, a petroleum refining process. Phenol is a toxic material; skin contact must be avoided.

phosphate ester--any of a group of *synthetic lubricants* having superior fire resistance. A phosphate ester generally has poor *hydrolytic stability*, poor compatibility with *mineral oil*, and a relatively low *viscosity index (V.I.)*. It is used as a fire-resistant *hydraulic fluid* in high-temperature applications.

phytotoxic--injurious to plants.

pig--solid plug inserted into pipelines and pushed through by fluid pressure. It may be used for separating two fluids being pumped through the line, or for cleaning foreign materials from a line.

pinion--see *gear*.

pipe still--see *distillation*.

piston sweep--see *sweep (of a piston)*.

plain bearing--see *bearing*.

plasticity--the property of an apparently solid material that enables it to be permanently deformed under the application of force, without rupture. (Plastic flow differs from fluid flow in that the *shear stress* must exceed a *yield point* before any flow occurs.)

plasticizer--any *organic compound* used in modifying plastics, *synthetic rubber*, and similar materials to incorporate flexibility and toughness.

plastisols and organosols--coating materials composed of *resins* suspended in a hydrocarbon liquid. An organosol is plastisol with an added solvent, which swells the resin particles, thereby increasing *viscosity*. Applications include spray coating, dipping, and coatings for aluminum, fabrics, and paper.

Platinum-Cobalt system--see *color scale*.

PNA (polynuclear aromatic)--any of numerous complex hydrocarbon compounds consisting of three or more *benzene* rings in a compact molecular arrangement. Some types of PNA's are known to be carcinogenic (cancer causing).

PNA's are formed in *fossil fuel* combustion and other heat processes, such a *catalytic cracking*. They can also form when foods or other organic substances are charred. PNA's occur naturally in many foods, including leafy vegetables, grain cereals, fruits, and meats.

poise--see *viscosity*.

polar compound--a chemical compound whose molecules exhibit electrically positive characteristics at one extremity and negative characteristics at the other. Polar compounds are used as additives in many petroleum products. Polarity gives certain molecules a strong affinity for solid surfaces; as lubricant additives (*oiliness agents*), such molecules plate out to form a tenacious, friction-reducing film. Some polar molecules are oil-soluble at one end and water-soluble at the other end; in lubricants, they act as *emulsifiers*, helping to form stable oil-water emulsions. Such

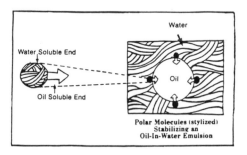

Polar Molecules (stylized)
Stabilizing an
Oil-In-Water Emulsion

lubricants are said to have good **metal-wetting** properties. Polar compounds with a strong attraction for solid contaminants act as *deter-gents* in engine oils by keeping contaminants finely dispersed.

pollutants   (atmospheric)--any substances released to the environment that threaten health or damage vegetation if present in sufficient concentration.   The major pollutants emitted as a result of man's industrial activity (largely through the combustion of *fossil fuels*) are: *sulfur oxides*--predominantly sulfur dioxide (SO₂)--*nitrogen oxides (NOx), carbon monoxide (CO), hydrocarbons (HC)*, and *particulates*. Such pollutants have a relatively short residence in the atmosphere before being removed by natural scavenging processes. SO₂ for example, has an atmospheric residence time of about four days.   There has thus been no evidence of a global buildup of these pollutants. In a given locality, however, pollutants can reach high concentrations in the atmosphere, causing respiratory ailments, as well as inhibiting growth of vegetation, turning soil acid, eroding masonry in buildings, and corroding metals.   See *emissions (automotive), emissions (stationary source)*.

polybutadiene rubber (BR)--one of the *stereo rubbers*, a term designating high uniformity of composition. High in abrasion resistance, BR is blended with *styrene-butadiene rubber (SBR)* for tire tread manufacture.   See *synthetic rubber*.

polyester--any of a number of synthetic *resins* usually produced by the polymerization of dibasic acid with a dihydric alcohol (see *polymer*).   Polyester resins have high sealing strength, and are weather resistant.   They are used in the manufacture of boat hulls, waterproof fibers, and adhesives.

polyethylene   (PE)--polymerized (see *polymer*) *ethylene*, ranging from a colorless liquid to a white solid; used in the manufacture of plastic films and sheets, and a wide variety of containers, kitchenware, tubing, etc.

polyglycols--*polymers* of ethylene or propylene oxides used as a *synthetic lubricant* base. Properties include very good *hydrolytic stability, high viscosity index (V.I.)*, and low *volatility*. Used particularly in water *emulsion* fluids.

polyisoprene rubber (IR)--one of the *stereo rubbers*, a term designating a high uniformity of composition. Sometimes called "synthetic natural rubber" because of its similar chemical composition, high tack, resiliency, and heat resistance. It can replace *natural rubber* in many applications.   See *synthetic rubber*.

polymer--substance formed by the linkage (polymerization) of two or more simple, unsaturated molecules (see *unsaturated hydrocarbon*), called **monomers**, to form a single heavier molecule having the same elements in the same proportions as the original monomers; i.e., each monomer retains its structural identity. A polymer may be liquid or solid; solid polymers may consist of millions of repeated linked units. A polymer made from two or more dissimilar monomers is called a **copolymer**; a copolymer composed of three different types of monomers is a **terpolymer**. *Natural rubber* and *synthetic rubbers* are polymers.

Major Monomers Used in Synthetic Rubbers

Monomer	Physical State	Chemical Structure
Butadiene	Gas	
Styrene	Liquid	
Isoprene	Liquid	
Isobutylene	Gas	
Ethylene	Gas	
Propylene	Gas	
Chloroprene	Liquid	
Acrylonitrile	Liquid	

Ⓒ Chlorine Atom
Ⓝ Nitrogen Atom

polymerization--in petroleum refining, polymerization refers to the combination of light, gaseous hydrocarbons, usually *olefins*, into high-molecular-weight hydrocarbons that are used in manufacturing motor gasoline and aviation fuel.

The product formed by combining two identical olefin molecules is called **a dimer**, and by three such molecules, a **trimer**. See *polymer*.

polynuclear aromatic--see *PNA*.

polyolefin--*polymer* derived by polymerization of relatively simple *olefins*. *Polyethylene* and *polyisoprene* are important polyolefins.

polyol ester--*synthetic lubricant* base, formed by reacting *fatty acids* with a polyol (such as a glycol) derived from petroleum. Properties include good *oxidation stability* at high temperatures and low *volatility*. Used in formulating lubricants for turbines, compressors, jet engines, and automotive engines.

polystyrene--hard, clear thermoplastic *polymer* of *styrene*, easily colored and molded for a variety of applications, including structural materials. It is a good thermal and electrical insulator and, in the form of expanded foam, extremely buoyant.

positive crankcase ventilation (PCV)--system for removing *blowby* gases from the crankcase and returning them, through the *carburetor* intake manifold, to the combustion chamber, where the recirculated hydrocarbons are burned, thus reducing hydrocarbon emissions to the atmosphere. A PCV valve, operated by engine vacuum, controls the flow of gases from the crankcase. PCV systems have been standard equipment in all U.S. cars since 1963, replacing the simpler vent, or breather, that allowed crankcase vapors to be emitted to the atmosphere.

pour point--lowest temperature at which an oil or *distillate* fuel is observed to flow, when cooled under conditions prescribed by test method ASTM D 97. The pour point is 3°C (5°F) above the temperature at which the oil in a test vessel shows no movement when the container is held horizontally for five seconds. Pour point is lower than *wax appearance point* or *cloud point*. It is an indicator of the ability of an oil or distillate fuel to flow at cold operating temperatures.

pour point depressant--*additive* used to lower the *pour point* of a petroleum product.

power--rate at which *energy* is used, or at which work is done. Power is commonly measured in terms of the watt (one joule per second) or horsepower (33,000 foot-pounds per minute, or 745.7 watts).

power factor--ratio of the power in watts (W) dissipated in an insulating medium to the product of the effective values of voltage (V) and current (I) in volt-amperes; a measure of the tendency of an insulating oil, which is a *dielectric* (non-conductor) of electricity), to permit leakage of current through the oil, as determined by test method ASTM D 924. Such current leakage is called **dielectric loss**. The lower the power factor, the lower the dielectric loss. Determination of power factor can be used to indicate not only the inherent dielectric properties of an oil, but the extent of deterioration of a used oil, since oxidation products and other polar contaminants reduce dielectric strength, causing the power factor to rise. Power factor is related to *dissipation factor*.

Powerforming®--catalytic *reforming* process patented by Exxon.

ppb--parts per billion.

ppm--parts per million.

pre-ignition--ignition of a fuel-air mixture in an *internal combustion engine* (gasoline) before the spark plug fires. It can be caused by a hot spot in the combustion chamber or a very high *compression ratio*. Pre-ignition reduces power and can damage the engine.

pressure--force per unit area, measured in kilopascals (kPa) or pounds per square inch (psi). Standard **atmospheric pressure** at sea level is 101.3 kPa (14.7 psi), or one *atmosphere*. **Gauge pressure,** as indicated by a conventional pressure gauge, is the pressure in excess of atmospheric pressure.

Absolute pressure is the sum of atmospheric and gauge pressures. Pressure is also expressed in terms of the height of a column of mercury that would exert the same pressure. One atmosphere is equal to 760 mm (29.9 in) of mercury.

pressure maintenance--method for increasing ultimate oil recovery by injecting gas, water, or other fluids into an oil *reservoir*, usually early in the life of the field in order to maintain or slow the decline of the reservoir pressures that force the oil to the surface.

pressure ratio (of a compressor)-- the ratio (r) of the absolute discharge pressure to the absolute pressure at the inlet. This is mathematically expressed as:

$$r = P_2/P_1$$

where P_2 is the discharge pressure and P_1 is the inlet pressure.

process oil--oil that serves as a temporary or permanent component of a manufactured product. *Aromatic* process oils have good solvency characteristics; their applications include proprietary chemical formulations, ink oils, and *extenders* in *synthetic rubbers*. Naphthenic (see *naphthene*) process oils are characterized by low *pour points* and good solvency properties; their applications include rubber compounding, printing inks, textile conditioning, leather tanning, shoe polish, rustproofing compounds, and dust suppressants. Paraffinic (see *paraffin*) process oils are characterized by low aromatic content and light color; their applications include furniture polishes, ink oils, and proprietary chemical formulations.

process stream--general term applied to a partially finished petroleum product moving from one refining stage to another; less commonly applied to a finished petroleum product. See *CAS Registry Numbers*.

propane--gaseous paraffinic hydrocarbon (C_3H^8) present in natural gas and crude oil; also termed, along with *butane*, liquefied petroleum gas (*LPG*). See *paraffin*.

propellant--volatile gas or liquid which, when permitted to escape from a pressurized container, carries with it particles or droplets of another material mixed or suspended in it. *Propane* and *butane* are common petroleum-derived propellants.

propylene--flammable gas (CH_3CHCH_2), derived from hydrocarbon *cracking*; used in the manufacture of polypropylene plastics.

psi--pounds per square inch.

psia--pounds per square inch absolute, equivalent to the gauge pressure plus atmospheric pressure. See *pressure*.

quality of steam--see *saturated steam*.

quenching--immersion of a heated manufactured steel part, such as a gear or axle, in a fluid to achieve rapid and uniform cooling. Petroleum oils are often used for this purpose. Quenching provides hardness superior to that possible if the heat-treated part were allowed to cool slowly in air. **Marquenching** is a slower cooling process that minimizes distortion and cracking. There are two types of marquenching; **martempering** and **austempering**; the latter is the slower process and helps improve ductility. See *quenching oil*.

quenching oil--also called **heat treating oil**; it is used to cool metal parts during their manufacture, and is often preferred to water because the oil's slower heat transfer lessens the possibility of cracking or warping of the metal. A quenching oil must have excellent *oxidation stability*, and should yield clean parts, essentially free of residue. See *quenching*. In refining terms, a quenching oil is an oil introduced into high temperature vapors of cracked (see *cracking*) petroleum fractions to cool them.

°R (Rankine)--see *saturated steam*.

°R (Reaumur)--see *temperature*

scales.

radical--atom or group of atoms with one or more unpaired electrons. A group of atoms functioning as a radical acts as a single atom, remaining intact during a chemical reaction.

raffinate--in *solvent extraction*, that portion of the oil which remains undissolved and is not removed by the selective solvent.

Ramsbottom carbon residue--see *carbon residue*.

R&O--rust- and-oxidation inhibited. A term applied to highly refined industrial lubricating oils formulated for long service in circulating systems, compressors, hydraulic systems, bearing housing, gear cases, etc. The finest R&O oils are often referred to as *turbine oils*.

Rankine (°R)--see *temperature scales*.

rapeseed oil (blown rapeseed oil)-- *fatty oil* used for compounding petroleum oil. See *compounded oil*.

rate of shear--see *shear rate*.

RC asphalt--see *cutback asphalt*.

reaction diluent--a material (usually a light *saturated hydrocarbon*, e.g., *pentane, hexane*) that is used as a *carrier* for the polymerization catalyst in the manufacture of *polyolefins* (see *polymer*). The material must be very pure, since impurities "poison" the catalyst or hinder the polymerization by reacting with the *olefins*.

Reaumur (°R)-- see *temperature scales*.

reclaimed aggregate material (RAM)--reprocessed pavement materials containing no reusable binding agent such as *asphalt*.

reclaimed asphalt pavement (RAP)--reprocessed pavement materials containing *asphalt* and aggregate (e.g., pebbles, crushed stone, shells).

recycling of asphalt paving--the reprocessing of old *asphalt* paving and associated materials for reuse as paving. There are three basic methods of recycling: 1) hot mix recycling, a process in which reclaimed asphalt and aggregate materials are combined with new asphalt, recycling agents (petroleum oils), and new aggregate (e.g., crushed stone) in a central plant to produce hot-mix paving mixtures; 2) cold mix recycling, the recombination of reclaimed asphalt and aggregate materials either in place, or at a central plant to produce a cold mix; 3) surface recycling, a process in which the old asphalt pavement surface is heated in place, scarified, remixed with new

asphalt or recycling agents as necessary, relaid, and rolled.

Redwood viscosity--method for determining the *viscosity* of petroleum products; it is widely used in Europe, but has limited use in the U.S. The method is similar to *Saybolt Universal viscosity*; viscosity values are reported as "Redwood seconds."

refined wax--low-oil-content wax, generally with an oil content of 1.0 mass percent or less, white in color, and meeting Food and Drug Administration standards for purity and safety. Refined waxes are suitable for the manufacture of drugs and cosmetics, for coating paper used in food packaging, and for other critical applications. Also called **fully refined wax**. See *wax (petroleum)*.

refining--series of processes for converting crude oil and its fractions to finished petroleum products. Following *distillation*, a petroleum fraction may undergo one or more additional steps to purify or modify it. These refining steps include: *thermal cracking, catalytic cracking, polymerization, alkylation, reforming, hydrocracking, hydroforming, hydrogenation, hydrogen treating, Hydrofining®, solvent extraction, dewaxing, de-oiling, acid treating, clay filtration,* and *deasphalting*. Refined lube oils may be blended with other lube stocks, and *addi-*

tives may be blended with *alkylates*, cracked stock or *reformates* to improve *octane number* and other properties of gasolines.

reformate--product of the *reforming* process.

refractive index--ratio of the velocity of light at a specified wave length in air to its velocity in a substance under examination. The refractive index of light-colored petroleum liquids can be determined by test method ASTM D 1218, using a refractometer and a monochromatic light source. Refractive index is an excellent test for uniform composition of *solvents*, rubber *process oils*, and other petroleum products. It may also be used in combination with other simple tests to estimate the distribution of naphthenic, paraffinic, (see *naphthene*, *paraffin*), and *aromatic* carbon atoms in a process oil.

refrigeration oil--lubricant for refrigeration compressors. It should be free of moisture to avoid reaction (hydrolysis) with halogenated refrigerants (see *halogen*) and prevent freezing of water particles that could impede refrigerant flow. It should have a low wax content to minimize wax precipitation on dilution with the refrigerant, which could block capillary-size passages in the circulating system.

Reid vapor pressure--see *vapor pressure*.

relative humidity--see *humidity*.

release agent--see *mold lubricant*.

Research Octane Number (RON) --see *octane number*.

reservoir--subsurface formation of porous, permeable rock (usually *sandstone, limestone,* or *dolomite*) containing oil or gas within the rock pores. A typical oil reservoir contains gas, oil, and water, which occupy the upper, middle, and lower regions of the reservoir, respectively. The flow of reservoir fluids from the rock to the *borehole* is driven by gas pressure or water pressure.

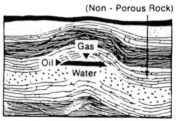

Oil and Gas Reservoir

residual fuel oil--see *fuel oil*.

residual odor--see *bulk odor*.

residuum--see *bottoms*.

resins--solid or semi-solid materials, light yellow to dark brown, composed of carbon, hydrogen, and oxygen. Resins occur naturally in plants, and are common in pines and firs, often appearing as globules on the bark. Synthetic resins, such as *polystyrene, polyesters,* and acrylics (see *acrylic resin*), are derived primarily from petroleum. Resins are widely used in the manufacture of lacquers, varnishes, plastics, adhesives, and rubber. See *plastisols and organosols.*

rheology--study of the deformation and flow of matter in terms of stress, strain, temperature, and time. The rheological properties of a grease are commonly measured by *penetration* and *apparent viscosity.*

rheopectic grease--grease that thickens, or hardens, upon being subjected to shear. The phenomenon is the opposite of *thixotropy.*

rich octane number--see *lean and rich octane numbers.*

ring oil--low-viscosity *R&O* oil for lubricating high-speed textile twister rings. It is normally light in color to prevent staining, and many are compounded with *fatty oils* to prevent wear under conditions of high-speed start-up. See *compounded oil.*

ring oiler--simple device for lubricating a *journal* bearing. A metal ring rides loosely on the journal shaft and the lowerpart of the ring dips into a small oil reservoir. The rotation of the shaft turns the ring, which carries oil up to the point of contact with the shaft and into the bearing. Though not ordinarily considered a *circulating lubrication system,* the ring oiler is similar in principle and generally requires a long-life oil of the *R&O* type.

ring-sticking--freezing of a piston ring in its groove, in a piston engine or reciprocating compressor, due to heavy deposits in the piston ring zone. This prevents proper action of the ring and tends to increase *blow-by* into the crankcase and to increase oil consumption by permitting oil to flow past the ring zone into the combustion chamber. See *engine deposits.*

Road Octane Number--see *octane number.*

road oil--a heavy petroleum oil, usually one of the slow-curing (SC) grades of liquid asphalt. See *cutback asphalt.*

rock oil--see *petroleum.*

rolling contact bearing--see *bearing.*

roll oil--oil used in hot- and cold-rolling of ferrous and non-ferrous metals to facilitate feed of the metal between the work rolls, improve the plastic deformation of the metal, conduct heat from the metal, and extend the life of the work rolls. Because of the pressures involved, a roll oil may be compounded (see *compounded oil*) or contain *EP additives*. In hot rolling, the oil may also be emulsifiable (see *emulsion*).

RON (Research Octane Number) --see *octane number*.

rotary bomb oxidation test--see *bomb oxidation stability*.

rotary drilling--drilling method utilizing a rotating bit, or cutting element, fastened to and rotated by a drill pipe, which also provides a passageway through which the drilling fluid or *mud* is circulated. Additional lengths of drill stem are added as drilling progresses.

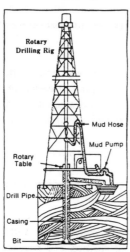

RS asphalt--see *emulsified anionic asphalt*.

rubber--see *natural rubber, synthetic rubber*.

rubber oil--any petroleum *process oil* used in the manufacture of rubber and rubber products. Rubber oils may be used either as rubber extender oils or as rubber process oils. **Rubber extender oils** are used by the synthetic rubber manufacturer to soften stiff elastomers and reduce their unit volume cost while improving performance characteristics of the rubber. **Rubber process oils** are used by the manufacturer of finished rubber products (tires, footwear, tubing, etc.) to speed mixing and compounding, modify the physical properties of the elastomer, and facilitate processing of the final product.

rubber oil classification--system of four-standard classifications for *rubber oils*, based on content of *saturated hydrocarbons, polar compounds*, and *asphaltenes*, as described by ASTM D 2226. The classifications are as follows:

| ASTM Type Oil | Hydrocarbon Type, mass % | | |
	Saturates	Polar Compounds	Asphaltenes
101 (highly aromatic)	20.0 max.	25 max.	0.75 max.
102 (aromatic)	20.1-35.0	12 max.	0.50 max.
103 (naphthenic)	35.1-65.0	6 max.	0.30 max.
104 naphthenic or paraffinic)	65.1 min.	1 max.	0.10 max.

Type 104 oils are subclassified into types 104A and 104B for *styrene-butadiene rubber* (SBR) only. Type 104A oils have a *viscosity-gravity constant* (VGC) greater than 0.820 (ASTM D 2501), and are naphthenic; Type 104B oils have a VGC of 0.820 max., and are paraffinic. See *aromatic, naphthene, paraffin*.

rubber swell--see *seal swell*.

rust inhibitor--type of *corrosion inhibitor* used in lubricants to protect the lubricated surfaces against rusting. See *R&O*.

rust preventive--compound for coating metal surfaces with a film that protects against rust; commonly used for the preservation of equipment in storage. The base material of a rust preventive may be a petroleum oil, solvent, wax, or asphalt, to which a *rust inhibitor* is added. A formulation consisting largely of a solvent and additives is commonly called a **thin-film rust preventive** because of the thin coating that remains after evaporation of the solvent. Rust preventives are formulated for a variety of conditions of exposure, e.g., short-time "in-process" protection, indoor storage, exposed outdoor storage, etc.

SAE (Society of Automotive Engineers)--organization responsible for the establishment of many U.S. automotive and aviation standards, including the viscosity classifications of engine oils and gear oils. See *SAE viscosity grades*.

SAE service classification--see *API Engine Service Categories*.

SAE viscosity grades--engine oil classification system developed by the Society of Automotive Engineers (SAE), based on the measured *viscosity* of the oil at either -18°C (-0.4°F), using test method

ASTM D 2602, or at 100°C (212°F), using ASTM D 445. If the viscosity is measured at -18°C, the grade number of the oil includes the suffix "W" (e.g., SAE 20W), denoting suitability for winter use. The grade number of an oil tested at 100°C is written without a suffix. It is common for an SAE 20 oil to meet the viscosity requirements at both -18°C and 100°C; such oils are designated 20W-20. Many oils have a high *viscosity index (V.I.)* and therefore may fall into more than one SAE grade classification; these are called *multi-grade oils*, and they are designated by two grade numbers indicating their lowest and highest classification (e.g., SAE 10W-40). The following table shows the viscosity range represented by each SAE grade:

SAE Viscosity Grade	centipoises (cp) at -18°C (ASTM D 2602) Max.	centistokes (cSt) at 100°C (ASTM D 445) Min.	Max.
5W	1,250	3.8	—
10W	2,500	4.1	—
20W⁽ᵃ⁾	10,000	5.6	—
20	—	5.6	Less than 9.3
30	—	9.3	Less than 12.5
40	—	12.5	Less than 16.3
50	—	16.3	Less than 21.9

⁽ᵃ⁾The designation SAE 15W may be used to identify SAE 20W oils that have a maximum viscosity of 5000 cp at -18°C.

After 1981, the low-temperature portion of the SAE viscosity classification system is to be superseded by a revised system, following an 18-month (1980-81) phase-in. The new system introduces the following key changes: 1) two new grades are added--a 0W grade for very low temperature (e.g., arctic) operation, and a 25W grade for temperate climates; 2) low-temperature viscosity is measured at a different temperature for each grade, rather than measuring all winter grades at -18°C; 3) a "borderline pumping temperature" is added to define the ability of a cold oil to flow to the oil pump inlet and through the pump discharge under cold-starting conditions. The revised SAE definitions for "W" grades are as follows:

SAE Viscosity Grade	Viscosity (cp)* at Temperature (°C)* Max	Borderline Pumping Temperature (°C)* Max	Viscosity (cSt)*** at 100°C Min	Max
0W	3250 at -30°	-35°	3.8	—
5W	3500 at -25°	-30°	3.8	—
10W	3500 at -20°	-25°	4.1	—
15W	3500 at -15°	-20°	5.6	—
20W	4500 at -10°	-15°	5.6	—
25W	6000 at - 5°	-10°	9.3	—

Revised Definitions for "W" Grades

*By proposed modification of ASTM D 2602. **ASTM D 3829. ***ASTM D 445

sandstone--sedimentary rock usually consisting of grains of quartz cemented by lime, silica, iron oxide, or other materials. Petroleum deposits are commonly found in sandstone formations. See *reservoir*.

saponification--process of converting certain chemicals into soaps, which are the metallic salts of organic acids. It is usually accomplished through reaction of a fat, *fatty acid*, or *ester* with an *alkali*-- an important process in *grease* manufacture.

saponification number--number of milligrams of potassium hydroxide (KOH) that combines with 1 gram of oil under conditions specified by test method ASTM D 94. Saponification number is an indication of the amount of fatty saponifiable material in a *compounded oil*. Caution must be used in interpreting test results if certain substances --such as sulfur compounds or *halogens*--are present in the oil, since these also react with KOH, thereby increasing the apparent saponification number.

saturated hydrocarbon--*hydrocarbon* with the basic formula $Cn H_2n+_2$; it is saturated with respect to hydrogen and cannot combine with the atoms of other elements without giving up hydrogen. Saturates are more chemically stable than *unsaturated hydrocarbons*.

saturated steam--the equilibrium condition at which the temperature of the steam is the same as that of the liquid water from which it is formed. Under this condition, steam containing no unvaporized water particles is called "100% quality." If the temperature of the steam is higher (at the same pressure), the steam is said to be "superheated." If, on the other hand, heat is removed from saturated steam (at the same pressure), as in a steam radiator, the quality drops, and the steam becomes wet, or condenses. At 50% quality, half of the weight represents water, the other half vapor. Saturated steam at atmospheric pressure has a temperature of 100°C (212°F).

Saybolt chromometer--see *color scale*.

Saybolt Furol viscosity--the efflux time in seconds required for 60 milliliters of a petroleum product to flow through the calibrated orifice of a Saybolt Furol *viscometer*, under carefully controlled temperature, as prescribed by test method ASTM D 88. The method differs from *Saybolt Universal viscosity* only in that the viscometer has a larger orifice to facilitate testing of very viscous oils, such as fuel oil (the word "Furol" is a contraction of "fuel and road oils"). The Saybolt Furol method has largely been supplanted by the *kinematic viscosity* method. See *viscosity*.

Saybolt Universal viscosity--the efflux time in seconds required for 60 milliliters of a petroleum product to flow through the calibrated orifice of a Saybolt Universal *viscometer*, under carefully controlled temperature, as prescribed by test method ASTM D 88. This method has largely been supplanted by the *kinematic viscosity* method. See *Saybolt Furol viscosity, viscosity*.

SBR--see *styrene-butadiene rubber*.

SC asphalt--see *cutback asphalt*.

scale wax--soft, semi-refined wax, distinguished from *slack wax* by having a generally lower oil content; usually derived from slack wax by extracting most of the oil from the wax. Used in candle manufacture, coating of carbon paper, and in rubber compounds to prevent surface cracking from sunlight exposure. See *oil content of petroleum wax, wax (petroleum)*.

scavenger--a component of lead *antiknock compounds* that reacts with the lead *radical* to form volatile lead compounds that can be easily scavenged from the engine through the exhaust system. Also, an individual who collects used lubricating oils for some secondary use.

scoring--distress marks on sliding metallic surfaces in the form of long, distinct scratches in the direction of motion. Scoring is an advanced stage of *scuffing*.

scrubber oil--see *absorber oil*.

scuffing--property of a wax coating that enables it to withstand abrasion. Scuff resistance is an indication of the extent to which paper-carton-coating machine operation affects the appearance of the coated carton. Poor scuff resistance of a wax can also cause excessive wax deposition on machine parts and adversely affect machine operations.

sealing strength--effectiveness of a coating wax in forming a tight, strong, heat-sealed package closure. See *laminating strength*.

seal oil--see *mineral seal oil*.

seal swell (rubber swell)--swelling of rubber (or other *elastomer*) *gaskets*, or seals, when exposed to petroleum, *synthetic lubricants*, or *hydraulic fluids*. Seal materials vary widely in their resistance to the effect of such fluids. Some seals are designed so that a moderate amount of swelling improves sealing action.

secondary production--see *secondary recovery*.

secondary recovery--restoration of an essentially depleted oil *reservoir* to production by injecting liquids or gases into the reservoir to flush out oil or to increase reservoir pressure. Also called **secondary production**. See *enhanced recovery, tertiary recovery*.

Series 3--obsolete specification for heavy-duty engine oils used in Caterpillar Tractor Company diesel engines. Caterpillar now specifies that the oil for its engines comply with Military Specification MIL-L-2104C or *API Engine Service Category* CD.

sett grease--any grease that changes from a fluid to a semifluid or plastic state after combination of the components, and often after packaging.

shale oil--see *oil shale*.

shear rate--rate at which adjacent layers of a fluid move with respect to each other, usually expressed as reciprocal seconds (also see *shear stress*). When the fluid is placed between two parallel surfaces moving relative to each other:

$$\frac{\text{shear}}{\text{rate}} = \frac{\text{relative velocity of surface (meters/ second)}}{\text{distance between surfaces (meters)}} = (\text{seconds})^{-1}$$

shear stress--frictional force overcome in sliding one "layer" of fluid along another, as in any fluid flow. The shear stress of a petroleum oil or other *Newtonian fluid* at a given temperature varies directly with *shear rate* (velocity). The ratio between shear stress and shear rate is constant; this ratio is termed *viscosity*. The higher the viscosity of a Newtonian fluid, the greater the shear stress as a function of rate of shear. In a *non-Newtonian fluid*--such as a *grease* or a *polymer*-containing oil (e.g., *multi-grade oil*)--shear stress is not proportional to the rate of shear. A non-Newtonian fluid may be said to have an *apparent viscosity*, a viscosity that holds only for the shear rate (and temperature) at which the viscosity is determined. See *Brookfield viscosity*.

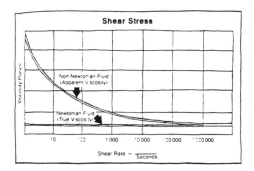

SI (Systeme International, International System of Units)--metric-based system of weights and measures adopted in 1960 by the 11th General Conference on Weights and Measures, in which 36 countries, including the U.S., participated.   SI consists of seven basic units:

Unit	Quantity
meter (m)	length
kilogram (kg)	mass
second (s)	time
ampere (A)	electric current
Kelvin (K)	thermodynamic temperature
mole (mol)	amount of substance
candela (cd)	luminous intensity

There are two supplemental units:

Unit	Quantity
radian (rad)	plane angle
steradian (sr)	solid angle

There are many derived units, each defined in terms of the base units; for example, the *newton (N)*--a unit of force--is defined by the formula kg x m/s^2, and the *joule (J)*, by the relationship Nxm.   See *metric system*.

sidestream--distillation   fraction taken from any level of a distillation tower other than as overhead or *bottoms*. See *distillation*.

sight fluid--transparent liquid in a sight-feed oiler through which the passage of the oil drops can be observed. The sight fluid must be immiscible with the oil. Water and glycerin are often used for this purpose.

silicate esters--class of *synthetic lubricants*, possessing good *thermal stability* and low *volatility*. Commonly used in military applications as high-temperature hydraulic fluids, weapons lubricants, and low-volatility greases.

silicone--generic term for a family of relatively inert liquid organosiloxane *polymers* used as *synthetic lubricants*; properties include high *viscosity index (V.I.)*, good high temperature *oxidation stability*, good *hydrolytic stability*, and low *volatility*. Silicones generally have poor *lubricity*, however. Applications include brake fluids, electric motors, oven and kiln preheater fans, automotive fans, plastic bearings, and electrical insulating fluids.

single-grade oil--engine oil that meets the requirements of a single SAE viscosity grade classification. See SAE *viscosity grades*.

slack wax--a semi-refined wax, distinguished from *scale wax* by having a generally higher oil content. Slack waxes with oil content below 10 mass percent are used for manufacture of religious candles, as a feedstock for chlorination processes, and in non-critical paper-making applications. Slack waxes with higher oil content are used in the manufacture of building materials, such as particle board. See *oil content of petroleum wax, wax (petroleum)*.

sleeve bearing--journal bearing. See *bearing*.

sludge--in gasoline engines, a soft, black, mayonnaise-like emulsion of water, other combustion by-products, and oil formed during low-temperature engine operation. Sludge plugs oil lines and screens, and accelerates wear of engine parts. Sludge deposits can be controlled with a *dispersant* additive that keeps the sludge constituents finely suspended in the oil. See *engine deposits*.

slumping--ability of grease to settle to the bottom of a container and form a level surface. When being pumped from the bottom of a container, grease must slump rapidly enough toward the pump suction to maintain flow.

slushing oil--non-drying petroleum-base *rust preventive* used in steel mills to protect the surfaces of steel sheets and strips after rolling.

smoke point (of aviation turbo fuel)--maximum flame height obtainable in a test lamp without causing smoking, as determined by

test method ASTM D 1322. It is a measure of burning characteristics; the higher the numerical rating, the cleaner burning the fuel.

SO_2, SO_3--see *sulfur oxide*.

soap thickener--see *grease*.

Society of Automotive Engineers--see *SAE*.

sodium soap grease--see *grease*.

SOD lead corrosion--test developed by Exxon to measure the corrosiveness of lubricating oils. A small lead panel of known weight and a copper panel (as catalyst) are attached to a spindle, which is immersed in a tube of the lubricant and rotated. Air is introduced at the bottom of the tube and allowed to bubble up through the sample. The weight loss by the lead panel after a specified period of time is a measure of the corrosivity of the oil.

softening point (asphalt)--temperature at which the harder asphalts used in applications other than paving reach an arbitrary degree of softening; usually determined by the ring and ball test method, ASTM D 36.

solids content--that portion of a protective coating material that remains on the surface after drying, often identified as **non-volatiles**.

soluble oil--emulsifiable *cutting fluid*.

solvent--compound with a strong capability to dissolve a given substance. The most common petroleum solvents are *mineral spirits, xylene, toluene, hexane, heptane,* and *naphthas*. *Aromatic*-type solvents have the highest solvency for organic chemical materials, followed by *naphthenes* and *paraffins*. In most applications the solvent disappears, usually by evaporation, after it has served its purpose. The evaporation rate of a solvent is very important in manufacture: rubber cements often require a fast-drying solvent, whereas rubber goods that must remain tacky during processing require a slower-drying solvent. Solvents have a wide variety of industrial applications, including the manufacture of paints, inks, cleaning products, adhesives, and *petrochemicals*. Other types of solvents have important applications in refining. See *solvent extraction*.

solvent extraction--refining process used to separate reactive components (*unsaturated hydrocarbons*) from lube *distillates* in order to improve the oil's *oxidation stability, viscosity index (V.I.),* and response to *additives*. Commonly used extraction media (solvents) are: *phenol*, N-methylpyrrolidone (NMP), *furfural*, liquid sulfur dioxide, and nitrobenzene. The oil and solvent are mixed in an extraction tower, resulting in the formation of two liquid phases: a heavy phase consisting of the undesirable unsaturates dissolved in the solvent, and a light phase consisting of high quality oil with some solvent dissolved in it. The phases are separated and the solvent recovered from each by distillation. The unsaturates portion, or extract, while undesirable in lubricating oils, is useful in other applications, such as rubber extender oils (see *rubber oil*) and *plasticizer* oils.

solvent neutral--high-quality *paraffin*-base oil refined by *solvent extraction*.

sour crude--crude oil containing appreciable quantities of *hydrogen sulfide* or other *sulfur* compounds, as contrasted to *sweet crude*.

spark-ignition engine --see *internal combustion engine*.

specific gravity--for petroleum products, the ratio of the mass of a given volume of product and the mass of an equal volume of water, at the same temperature. The standard reference temperature is 15.6°C (60°F). Specific gravity is determined by test method ASTM D 1298: the higher the specific gravity, the

Hydrometer

heavier the product. Specific gravity of a liquid can be determined by means of a **hydrometer**, a graduated float weighted at one end, which provides a direct reading of specific gravity, depending on the depth to which it sinks in the liquid.   A related measurement is **density**, an absolute unit defined as mass per unit volume--usually expressed as kilograms per cubic meter (kg/m^3). Petroleum products may also be defined in terms of **API gravity** (also determinable by ASTM D 1298), in accordance with the formula:

$$\text{API gravity (degree)} = \frac{141.5}{\text{specific gravity (at 15.6/15.6°C)}} - 131.5$$

Hence, the higher the API gravity value, the lighter the material, or the lower its specific gravity.

specific heat--ratio of the quantity of heat required to raise the temperature of a substance one degree Celsius (or Fahrenheit) and the heat required to raise an equal mass of water one degree.

speed factor--see *dN factor*.

spin finish--see *fiber lubricant*.

spindle oil--low-*viscosity* oil of high quality for the lubrication of high-speed textile and metal-working (grinding) machine spindles. In addition to the rust and oxidation (*R&O*) inhibitors needed for prolonged service in humid environments, spindle oils are often fortified with anti-wear agents to reduce torque load and wear, especially at start-up.

spindle test--test to determine the performance life of a grease. Under test method ASTM D 3336, a grease-lubricated SAE No. 204-size ball bearing on a spindle, or shaft, is rotated at 10,000 rpm under light loads and elevated temperatures. The test continues until bearing failure or until completion of a specified number of hours of running time.

spray oil--see *orchard spray oil*.

spur gear--see *gear*.

squeeze lubrication--phenomenon that occurs when surfaces, such as two gear teeth, move toward each other rapidly enough to develop fluid pressure within the lubricant that will support a load of short duration. The lubricant's viscosity prevents it from immediately flowing away from the area of contact.

SS asphalt--see *emulsified anionic asphalt*.

SSF (Saybolt Furol seconds)--see *Saybolt Furol viscosity*.

SSU, SUS (Saybolt Universal seconds)--see *Saybolt Universal viscosity*.

stationary source emissions--see *emissions (stationary source)*.

steam cylinder oil--see *cylinder oil.*

steam turbine--see *turbine.*

stereo rubber--*elastomer* with a highly uniform arrangement of repeating molecular units (stereo-isomers) in its structure. *Polybutadiene rubber* and *polyisoprene rubber* are stereo rubbers.

stick-slip motion--erratic, noisy motion characteristic of some machine *ways*, due to the starting friction encountered by a machine part at each end of its back-and-forth (reciprocating) movement. This undesirable effect can be overcome with a way lubricant, which reduces starting friction.

Stoddard solvent--*mineral spirits* with a minimum *flash point* of 37.8°C (100°F), relatively low odor level, and other properties conforming to Stoddard solvent specifications, as described in test method ASTM D 484. Though formulated to meet dry cleaning requirements, Stoddard solvents are widely used wherever this type of mineral spirits is suitable.

stoichiometric--the exact proportion of two or more substances that will permit a chemical reaction with none of the individual reactants left over. See *combustion.*

stoke--see *viscosity.*

straight-chain paraffin--see *normal paraffin.*

straight mineral oil--petroleum oil containing no *additives.* Straight mineral oils include such diverse products as low-cost once-through lubricants (see *once-through lubrication*) and thoroughly refined *white oils.* Most high-quality lubricants, however, do contain additives. See *mineral oil.*

straight-tooth gear--see *gear.*

strike through--undesirable migration of wax or oil through a paper substrate. A commonly used term in paper laminating operations, but also encountered in partial impregnation of corrugated board with wax. In the printing industry, the term refers to migration of printing ink, formulated with oil or solvent, to the reverse side of the web before setting.

strong acid number--see *neutralization number.*

strong base number--see *neutralization number.*

structural stability--resistance of a grease to change in *consistency* when severely worked in service. Also called **shear stability** and **mechanical stability**.

styrene--colorless liquid (C_8H_8) used as the monomer (see *polymer*) for *polystyrene* and *styrene-butadiene rubber*.

styrene-butadiene rubber (SBR)--general-purpose *synthetic rubber* with good abrasion resistance and tensile properties. SBR can be greatly extended with oil without degrading quality. Applications include automobile tires and wire insulation.

sulfonate--hydrocarbon in which a hydrogen atom has been replaced with the highly polar (SO_2OX) group, where X is a metallic ion or *alkyl* radical. Petroleum sulfonates are refinery by-products of the sulfuric acid treatment of *white oils*. Sulfonates have important applications as *emulsifiers* and chemical intermediates in *petrochemical* manufacture. Synthetic sulfonates can be manufactured from special *feedstocks* rather than from white oil *base stocks*. See *polar compound*.

sulfur--common natural constituent of petroleum and petroleum products. While certain sulfur compounds are commonly used to improve the EP, or load-carrying, properties of an oil (see *EP oil*), high sulfur content in a petroleum product may be undesirable as it can be corrosive and create an environmental hazard when burned (see *sulfur oxide*). For these reasons, sulfur limitations are specified in the quality control of fuels,

solvents, etc. Sulfur content can be determined by *ASTM* tests.

sulfur oxide--major atmospheric pollutant, predominantly sulfur dioxide (SO_2) with some sulfur trioxide (SO_3), primarily emitted from stationary combustion sources (furnaces and boilers). Sulfur oxides are formed whenever fuels containing sulfur are burned. SO_2 is also present in the air from natural land and marine fermentation processes. See *emissions (stationary source)*, *pollutants*.

supercharger--device utilizing a blower or pump to provide intake air to the *carburetor* of an *internal combustion engine* at pressures above atmospheric. Supercharging provides a greater air charge to the cylinders at high crankshaft speeds and at high altitudes, thereby boosting engine power without increasing engine size. Because supercharging maintains maximum intake charge, it offers particular advantages at high altitudes, where the atmosphere contains less oxygen. Some supercharger systems utilize aftercooling to further increase the density of the charge. The blower may be geared to the crankshaft or, in the case of the **turbocharger**, it may consist of a turbine driven by the exhaust gases to operate the centrifugal blower. See *volumetric efficiency*.

supertanker--oil tanker with capacity over 100,000 deadweight tons (dwt); also called **Very Large Crude Carrier** (VLCC). A supertanker with a capacity over 500,00 dwt is called an **Ultra Large Crude Carrier** (ULCC).

surface-active agent--chemical compound that reduces interfacial tension between oil and water and is thus useful as an *emulsifier* in cutting oils (see *cutting fluid*). Sodium *sulfonates* or soaps of *fatty acids* are commonly used for this purpose.

surface ignition--see *afterrunning*.

surfactant--surface-active agent that reduces interfacial tension of a liquid. A surfactant used in a petroleum oil may increase the oil's affinity for metals and other materials.

susceptibility--the tendency of a gasoline toward an increase in *octane number* by addition of a specific amount of a particular *lead alkyl* antiknock compound. See *antiknock compound*.

sweep (of a piston)--internal cylinder surface area over which a piston of a reciprocating compressor moves during its stroke. Total piston sweep is a consideration in the determination of oil-feed rates for some reciprocating compressor cylinders, and may be determined as:

length of stroke x cylinder circumference diameter x 2 x no. of cylinders x rpm x minutes of operation

sweet crude--crude oil containing little or no *sulfur*. See *sour crude*.

syneresis--loss of liquid component from a lubricating grease caused by shrinkage or rearrangement of the structure due to either physical or chemical changes in the thickener; a form of *bleeding*.

synlube--see *synthetic lubricant*.

synthetic gas--see *synthetic oil and gas*.

synthetic lubricant--lubricating fluid made by chemically reacting materials of a specific chemical composition to produce a compound with planned and predictable properties; the resulting *base stock* may be supplemented with *additives* to improve specific properties. Many synthetic lubricants--also called **synlubes**--are derived wholly or primarily from *petrochemicals*; other synlube raw materials are derived from coal and oil shale, or are lipochemicals (from animal and vegetable oils). Synthetic lubricants may be superior to petroleum oils in specific performance areas. Many exhibit higher *viscosity index (V.I.)*, better *thermal stability* and *oxidation stability*, and low *volatility* (which reduces oil consumption). Individual synthetic lubricants offer specific outstanding properties:

phosphate esters, for example, are fire resistant, *diesters* have good oxidation stability and *lubricity*, and *silicones* offer exceptionally high V.I. Most synthetic lubricants can be converted to grease by adding thickeners.   Because synthetic lubricants are higher in cost than petroleum oils, they are used selectively where performance or safety requirements may exceed the capabilities of a conventional oil. The following table lists the principal classes of synthetic lubricants:

alkylated aromatics[1]	polyol esters[2]	silicones
olefin oligomers[1]	polyglycols	silicate esters
dibasic acid esters[2]	phosphate esters	halogenated
		hydrocarbons
[1]organic hydrocarbon	[2]organic ester	

synthetic oil and gas--any oil or gas suitable as fuel, but not produced by the conventional means of pumping from underground reserves.   Synthetic gas may be derived from coal, *naphtha*, or liquid petroleum gas (*LPG*); and synthetic oil may be derived from coal, *oil shale*, and tar sands.

synthetic rubber--any *petrochemical*-based *elastomer*. Like *natural rubber*, synthetic rubbers are *polymers*, consisting of a series of simple molecules, called monomers, linked together to form large chain-like molecules.   The chain forms a loose coil that returns to its coiled form after it is extended. See under individual listings: *butyl rubber, ethylene-propylene rubber, natural rubber, neoprene rubber,*

nitrile rubber, polybutadiene rubber, polyisoprene rubber, styrene-butadiene rubber.

synthetic turbo oil--non-petroleum lubricant for aircraft gas *turbines* generally made from an *ester* base. It is characterized by high *oxidation stability* and *thermal stability*, good load-carrying capacity, and the extreme low *volatility* required to prevent excessive evaporation under wide operating-temperature conditions.

Systeme International--see *SI*.

systemic effect--toxic effect that is produced in any of the organs of the body after a toxicant has been absorbed into the bloodstream.

tackiness agent--*additive* used to increase the adhesive properties of a lubricant, improve retention, and prevent dripping and splattering.

Tag closed tester--apparatus for determining the *flash point* of petroleum liquids having a *viscosity* below 5.8 centistokes (cSt) at 37.8°C (100°F) and a flash point below 93°C (200°F), under test methods prescribed in ASTM D 56. The test sample is heated in a closed cup at a specified constant rate. A small flame of specified size is introduced into the cup through a shuttered opening at specified intervals.   The lowest temperature at which the vapors above the sample briefly ignite is the flash point. See *Pensky-Martens closed tester*.

Tag open cup--apparatus for determining the *flash point* of hydrocarbon liquids, usually *solvents*, having flash points between -17.8° and 168°C (0° to 325°F), under test methods prescribed in ASTM D 1310. The test sample is heated in an open cup at a slow, constant rate. A small flame is passed over the cup at specified intervals. The lowest temperature at which the vapors above the sample briefly ignite is the flash point. See *Cleveland open cup*.

Tag-Robinson colorimeter--see *color scale*.

tail-end volatility--see *distillation test*.

technical white oil--see *white oil*.

TEL (tetraethyl lead)--see *lead alkyl*.

tempering--hardening or strengthening of metal by application of heat or by alternate heating and cooling.

tempering oil--oil used as a medium for heating metals to their *tempering* temperature to relieve stress and improve toughness and ductility.

temperature scales--arbitrary thermometric calibrations that serve as convenient references for temperature determination. There are two thermometric scales based on the freezing and boiling point of water at a pressure of one **atmosphere**: the **Fahrenheit** (F) scale (32° = freezing, 212° = boiling)) and the **Celsius** (C), or **Centigrade**, scale (0° = freezing, 100° = boiling). Additionally, there are two scales in which 0° = absolute zero, the temperature at which all molecular movement theoretically ceases: the **Kelvin** (K), or **Absolute** (°A), scale and the **Rankine** (°R) scale, which are related to the Celsius and the Fahrenheit scales, respectively (0°K = -273.16°C; 0°R = -459.69°F). The four scales can be related to each other by the following formulas:

$$°C = 5/9 \ (°F–32) \quad °F = 9/5 \ °C + 32$$
$$°K = °C + 273.16 \quad °R = °F + 459.69$$

Another scale based on the thermometric properties of water is the **Reaumur** scale, in which the freezing point is set at zero degrees and the boiling point at 80 degrees. This scale has only limited application.

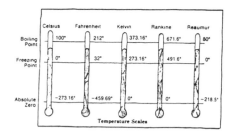

Temperature Scales

terpolymer--*copolymer* formed by the polymerization of three different *monomers*. An example of a terpolymer is *EPDM rubber*, made from *ethylene, propylene,* and a third monomer (usually a *diolefin*). See *polymer.*

tertiary recovery--any method employed to increase removal of hydrocarbons from a *reservoir* after *secondary recovery* methods have been applied.

tetraethyl lead (TEL)--see *lead alkyl.*

tetramethyl lead (TML)--see *lead alkyl.*

texture--that property of a lubricating grease which is observed when a small portion of it is compressed and the pressure slowly released. Texture should be described in the following terms: **brittle**--has a tendency to rupture or crumble when separated; **buttery**--separates in short peaks with no visible fibers; **long fiber**--shows tendency to stretch or string out into bundles of fibers; **short fiber**--shows short break-off with evidence of fibers; **resilient**--capable of withstanding moderate compression without permanent deformation or rupture; **stringy**--shows tendency to stretch or string out into long fine threads, but with no visible evidence of fiber structure.

thermal cracking--in refining, the breaking down of large, high-boiling hydrocarbon molecules into smaller molecules in the presence of heat and pressure. See *cracking.*

thermal stability--ability to resist chemical degradation at high temperatures.

thermal value--see *heat of combustion.*

thin-film rust preventive--see *rust preventive.*

thixotropy--tendency of grease or other material to soften or flow when subjected to shearing action. Grease will usually return to its normal consistency when the action stops. The phenomenon is the opposite of that which occurs with *rheopectic grease.* Thixotropy is also an important characteristic of *drilling fluids,* which must thicken when not in motion so that the cuttings in the fluid will remain in suspension.

threshold limit value (TLV)--see *occupational exposure limit.*

throttle plate--see *carburetor.*

time-weighted average (TWA)-- atmospheric concentration of a substance, in parts per million by volume, measured over a seven or eight-hour workday and 40-hour week.

Timken EP test--measure of the extreme-pressure properties of a lubricating oil (see *EP oil*). The test utilizes a Timken machine, which consists of a stationary block pushed upward, by means of a lever arm system, against the rotating outer race of a roller bearing, which is lubricated by the product under test. The test continues under increasing load (pressure) until a measurable wear scar is formed on the block. Timken OK load is the heaviest load that a lubricant can withstand before the block is scored (see *scoring*).

TLV (threshold limit value)--see *occupational exposure limit*.

TML (tetramethyl lead)--see *lead alkyl*.

toluene--aromatic hydrocarbon ($C_6H_5CH_3$) with good solvent properties; used in the manufacture of lacquers and other industrial coatings, adhesives, printing ink, insecticides, and chemical raw materials. Also called **toluol**.

toluene

toluene insolubles--see *insolubles*.

toluol--see *toluene*.

torque fluid--lubricating and power-transfer medium for commercial automotive torque converters and transmissions. It possesses the low viscosity necessary for torque transmission, the lubricating properties required for associated gear assemblies, and compatibility with seal materials.

total acid number--see *neutralization number*.

total base number--see *neutralization number*.

transformer oil--see *insulating oil*.

tribology--science of the interactions between surfaces moving relative to each other. Such interactions usually involve the interplay of two primary factors: the load, or force, perpendicular to the surfaces, and the frictional force that impedes movement. Tribological research on friction reduction has important energy conservation applications, since friction increases energy consumption. See *friction*.

trimer--see *polymerization*.

turbine--device that converts the force of a gas or liquid moving across a set of rotor and fixed blades into rotary motion. There are three basic types of turbines: gas, steam, and hydraulic. **Gas turbines** are powered by the expansion of compressed gases generated by the combustion of a fuel. (See *internal combustion engine*). Some of the power thus produced is used to drive an air compressor, which provides the air necessary for combustion of the fuel. In a turbo-jet aircraft engine, the turbine's only function is to drive the compressor: the plan is propelled by the force of the expanding gases escaping from the rear of the engine. In other applications, however, the rotor shaft provides the driving thrust to some other mechanism, such as a propeller or generator. Thus, gas turbines power not only turbo-jet aircraft, but also turbo-prop aircraft, locomotives, ships, compressors, and small-to-medium-size electric utility generators. Gas turbine-powered aircraft present severe lubrication demands that are best met with a *synthetic turbo oil*. **Steam turbines** employ steam that enters the turbine at high temperature and pressure and expands across both rotating and fixed blades (the latter serving to direct the steam). Steam turbines, which power large electric generators, produce most of the world's electricity. Only the highest-quality lubricants are able to withstand the wet conditions and high temperatures associated with steam turbine operation. The term *turbine oil* has thus become synonymous with quality. **Hydraulic turbines** (water turbines) are either impulse type, in which falling water hits blades or buckets on the periphery of a wheel that turns a shaft, or reaction type, where water under pressure emerg-

es from nozzles on the wheel, causing it to turn. Hydraulic turbines can be used to produce electric power near reservoir or river dams.

turbine oil--top-quality rust- and oxidation-inhibited (*R&O*) oil that meets the rigid requirements traditionally imposed on steam-turbine lubrication. (See *turbine*.) Quality turbine oils are also distinguished by good *demulsibility*, a requisite of effective oil-water separation. Turbine oils are widely used in other exacting applications for which long service life and dependable lubrication are mandatory. Such applications include circulating systems, compressors, hydraulic systems, gear drives, and other equipment. Turbine oils can also be used as *heat transfer fluids* in open systems, where *oxidation stability* is of primary importance.

turbocharger--see *supercharger*.

turbo fuel (jet fuel)--*kerosene*-type fuel used in gas-turbine-powered aircraft (see *turbine*). Important properties of a turbo fuel include: clean and efficient burning, ability to provide adequate energy for thrust, resistance to chemical degradation in storage or when used as a heat transfer medium on the aircraft, non-corrosiveness, ability to be transferred and metered under all conditions, volatility high enough for burning but low enough to prevent excessive losses from tank vents, and freedom from dirt, rust, water, and other contaminants. Turbo fuel has different properties than *aviation gasoline*, which fuels piston-engine air-craft.

turbo oil--see *synthetic turbo oil*.

TWA--see *time-weighted average*.

two-stroke cycle--see *internal combustion engine*.

Ultra Large Crude Carrier (ULCC)--see *supertanker*.

ultraviolet absorbance--measurement of the ultraviolet absorption of petroleum products, determined by standardized tests, such as ASTM D 2008. *Aromatics* absorb more ultraviolet light than do *naphthenes* and *paraffins*, and the amount of absorbance can be used as an indication of the amount of aromatics in a product. Certain polynuclear aromatics (*PNA's*) are known carcinogens (cancer-causing substances), with peaks of absorbance generally between 280 and 400 millimicrons. The Food and Drug Administration (FDA) has therefore imposed limits on the amount of ultraviolet absorbance at these wavelengths for materials classified as food additives. However, not all materials with ultraviolet absorbance at these wavelengths are carcinogenic.

undisturbed penetration--see *penetration (grease)*.

United States Geological Survey--see *USGS*.

United States Pharmacopeia--see *USP*.

unleaded gasoline--gasoline that derives its *antiknock* properties from high-octane hydrocarbons or from non-lead *antiknock compounds*, rather than from a lead additive. See *lead alkyl*.

unsaturated hydrocarbon--hydrocarbon lacking a full complement of hydrogen atoms, and thus characterized by one or more double or triple bonds between carbon atoms. Hydrocarbons having only one double bond between adjacent carbon atoms in the molecule are called *olefins*; those having two double bonds in the molecule are *diolefins*. Hydrocarbons having alternating single and double bonds between adjacent carbon atoms in a benzene-ring configuration are called *aromatics*. Hydrocarbons with a triple bond between carbon atoms are called *acetylenes*. Unsaturated hydrocarbons readily attract additional hydrogen, oxygen, or other atoms, and are therefore highly reactive. See *hydrocarbon, saturated hydrocarbon, hydrogenation*.

unsulfonated residue--measure of the volume of unsulfonated residue in plant spray oils of petroleum origin, as determined by test method ASTM D 483. Unsulfonated residue consists of those components of the oil that will not react with concentrated sulfuric acid. This determination is useful for distinguishing between oils suitable for various types of spraying applications.   An oil with a high unsulfonated residue (92 volume percent minimum) is required for spraying orchard crops in the leaf or bud stage.   See *orchard spray oil*.

unworked penetration--see *penetration (grease)*.

USGS (United States Geological Survey)--bureau of the Department of the Interior, responsible for performing surveys and investigations covering U. S. topography, geology, and mineral and water resources and for enforcing departmental regulations applicable to oil, gas, and other mining activities.

USP (United States Pharmacopeia)--compendium of drugs, drug formulas, quality standards and tests published by the United States Pharmacopeial Convention, Inc., which also publishes the *NF* (National Formulary). The purpose of the USP is to ensure drug uniformity and to maintain and upgrade standards of drug quality and purity, as well as establish packaging, labeling, and storage requirements. The USP includes standards for *white oils* under two classifications: "Mineral Oil" for heavy grades, and "Mineral Oil Light" for lighter

grades.

vacuum tower--see *distillation*.

valve beat-in--wear on the valve face or valve seat in *internal combustion engines* resulting from the pounding of the valve on the seat. Also called **valve sink** or **valve recession**.

vapor lock--disruption of fuel movement to a *gasoline* engine *carburetor* caused by excessive vaporization of gasoline. Vapor lock occurs when the fuel pump, which is designed to pump liquid, loses suction as it tries to pump fuel vapor. The engine will usually stall, but in less severe cases may accelerate sluggishly or *knock* due to an excessively lean fuel mixture. Automotive engines are more likely to experience vapor lock during an acceleration that follows a short shutdown period.   Vapor lock problems are most likely to occur in the late spring on unseasonably warm days, before the more volatile winter grades of gasoline have been replaced by the less volatile spring and summer grades (see *volatility*). Vapor lock can also occur in other types of pumping systems where volatile liquids are being handled.

vapor pressure--pressure of a confined vapor in equilibrium with its liquid at a specified temperature; thus, a measure of a liquid's *volatility*. Vapor pressure of gasoline and other volatile petroleum products is commonly measured in accordance with test method ASTM D 323 (**Reid vapor pressure**). The apparatus is essentially a double-chambered bomb. One chamber, fitted with a pressure gauge, contains air at atmospheric pressure; the other chamber is filled with the liquid sample. The bomb is immersed in a 37.8°C (100°F) bath, and the resulting vapor pressure of the sample is recorded in pounds per square inch (psi). Reid vapor pressure is useful in predicting seasonal gasoline performance (e.g., higher volatility is needed in cold weather, and lower volatility in hot weather), as well as the tendencies of gasolines, solvents, and other volatile petroleum products toward evaporative loss and fire hazard.

varnish--hard, varnish-like coating formed from oil oxidation products, that bakes on to pistons during high-temperature automotive engine operation. Varnish can accelerate cylinder wear. Varnish formation can be reduced with the use of a *detergent-dispersant* and an *oxidation inhibitor* in the oil. See *engine deposits*.

Very Large Crude Carrier (VLCC)--see *supertanker*.

V.I.--see *viscosity index (V.I.)*.

V.I. (viscosity index) improver--see *viscosity index (V.I.) improver*.

viscometer--device for measuring *viscosity*; commonly in the form of a calibrated capillary tube through which a liquid is allowed to pass at a controlled temperature in a specified time period. See *kinematic viscosity*.

viscosity--measurement of a fluid's resistance to flow. The common metric unit of *absolute viscosity* is the **poise**, which is defined as the force in dynes required to move a surface at a speed of one centimeter per second, with the surfaces separated by a fluid film one centimeter thick. For convenience, the **centipose** (cp)--one one-hundredth of a poise--is the unit customarily used. Laboratory measurements of viscosity normally use the force of gravity to produce flow through a capillary tube (*viscometer*) at a controlled temperature. This measurement is called *kinematic viscosity*. The unit of kinematic viscosity is the **stoke**, expressed in square centimeters per second. The more customary unit is the **centistoke** (cSt)--one one-hundredth of a stoke. Kinematic viscosity can be related to absolute viscosity by the equation:

$$cSt = cp \div \text{fluid density}.$$

In addition to kinematic viscosity, there are other methods for determining viscosity, including *Saybolt Universal viscosity, Saybolt Durol viscosity, Engler viscosity*, and *Redwood viscosity*. Since viscosity varies inversely with temperature, its value is meaningless unless the temperature at which it is determined is reported.

viscosity (asphalt)--determined by any of several *ASTM* test methods. Two common methods are ASTM D 2170 and D 2171. The former method measures *kinematic viscosity*, that is, viscosity under the force of gravity, by allowing a test sample to flow down a capillary tube (*viscometer*) at a temperature of 135°C (275°F); the viscosity is expressed in centistokes. The latter method measures *absolute viscosity*. The liquid, at a temperature of 60°C (140°F), is drawn up a tube by applying a vacuum; the viscosity is expressed in poises. In both tests the viscosity is obtained by multiplying the flow time in seconds by the viscometer calibration factor. See *penetration (asphalt), viscosity*.

viscosity grading (asphalt)--classification system for *asphalt cement*, defined in American Association of State Highway Transportation Officials (AASHTO) Specification M 226, and based on tests for *viscosity, flash point*, ductility, purity, etc., as outlined in test method ASTM D 3381. There are five standard grades, from softest to hardest: AC-2.5, AC-5, AC-10, AC-20, AC-40. Asphalt cement is also classified by *penetration grading*. See *viscosity (asphalt)*.

viscosity-gravity constant (VGC)-- indicator of the approximate hydrocarbon composition of a petroleum oil. As described in test method ASTM D 2501, VGC is calculated from one of the following equations, depending on the temperature at which viscosity is determined (VGC at 37.8°C [100°F] is the preferred equation):

$$\text{VGC @ 37.8°C (100°F)} = \frac{10G - 1.0752 \log (V - 38)}{10 - \log (V - 38)}$$

where: G = specific gravity @ 15.6°/15.6°C [60/60°F].
and
V = Saybolt Universal viscosity @ 37.8°C (100°F).

or

$$\text{VGC @ 98.9°C (210°F)} = \frac{G - 0.1244 \log (V_1 - 31)}{0.9255 - 0.0979 \log (V_1 - 31)} - 0.0839$$

where: G = specific gravity at 15.6°/15.6°C (60/60°F). and
V_1 = Saybolt Universal viscosity at 98.9° (210°F).

Values of VGC near 0.800 indicate an oil of paraffinic character (see *paraffin*); values close to 1.00 indicate a preponderance of *aromatic* structures. Like other indicators of hydrocarbon composition (as opposed to a specific laboratory analysis), VGC should not be indiscriminately applied to residual oils (see *bottoms*), asphaltic materials, or samples containing appreciable quantities of non-hydrocarbons. See *Saybolt Universal Viscosity, specific gravity*.

viscosity index (V.I.)--empirical, unitless number indicating the effect of temperature change on the *kinematic viscosity* of an oil. Liquids change *viscosity* with temperature, becoming less viscous when heated; the higher the V. I. of an oil, the lower its tendency to change viscosity with temperature. The V.I. of an oil--with known viscosity at 40°C and 100°C--is determined by comparing the oil with two standard oils having an arbitrary V.I. of 0 and 100, respectively, and both having the same viscosity at 100°C as the test oil. The following formula is used, in accordance with test method ASTM D 2270:

$$V.I. = \frac{L-U}{L-H} \times 100$$

where L is the viscosity at 40°C of the 0-V.I. oil, H is the viscosity at 40°C of the 100-V.I. oil, and U is the viscosity at 40°C of the test oil. There is an alternative calculation, also in ASTM D 2270, for oils with V.I.'s above 100. The V.I. of paraffinic oils (see *paraffin*) is inherently high, but is low in naphthenic oils (see *naphthene*), and even lower in *aromatic* oils (often below 0). The V.I. of any petroleum oil can be increased by adding a *viscosity index improver*. High-V.I. lubricants are needed wherever relatively constant viscosity is required at widely varying temperatures. In an automobile, for example, an engine oil must flow freely enough to permit cold starting, but must be viscous enough after warm-up to provide full lubrication. Similarly, in an aircraft hydraulic system, which

may be exposed to temperatures about 38°C at ground level and temperatures below -54°C at high altitudes, consistent hydraulic fluid performance requires a high viscosity index.

viscosity index (V.I.) improver-- lubricant *additive*, usually a high-molecular-weight *polymer*, that reduces the tendency of an oil to change *viscosity* with temperature. *Multi-grade oils*, which provide effective lubrication over a broad temperature range, usually contain V.I. improvers. See *viscosity index*.

viscosity-temperature relation-ship--the manner in which the *viscosity* of a given fluid varies inversely with temperature. Because of the mathematical relationship that exists between these two variables, it is possible to predict graphically the viscosity of a petroleum fluid at any temperature within a limited range if the viscosities at two other temperatures are known. The charts used for this purpose are the *ASTM* Standard Viscosity-Temperature Charts for Liquid Petroleum Products, available in 6 ranges. If two known

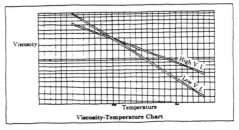

Viscosity-Temperature Chart

viscosity-temperature points of a fluid are located on the chart and a straight line drawn through them, other viscosity-temperature values of the fluid will fall on this line; however, values near or below the *cloud point* of the oil may deviate from the straight-line relationship.

VM&P Naphtha--Varnish Makers and Painters Naphtha: term for a *naphtha* commonly used as a solvent in paints and varnishes.

volatility--expression of evaporation tendency. The more volatile a petroleum liquid, the lower its boiling point and the greater its flammability. The volatility of a petroleum product can be precisely determined by tests for evaporation rate; also, it can be estimated by tests for *flash point* and *vapor pressure*, and by *distillation tests*.

volumetric efficiency--ratio of the weight of air drawn into the cylinder of an operating *internal combustion engine* to the weight of air the cylinder could hold at rest when the piston is at the bottom of the stroke and the valves are fully closed. Any restriction of air flow into the cylinder reduces volumetric efficiency, which, in turn, reduces power output. The volumetric efficiency of an automotive engine is usually slightly more than 80% at about half the rated speed of the engine, then decreases considerably at higher speed, thus limiting the power output of the engine. The air charge to the cylinder can be increased at high speeds by means of supercharging. See *supercharger*.

wash oil--see *absorber oil*.

wax (petroleum)--any of a range of relatively high-molecular-weight hydrocarbons (approximately C_{16} to C_{50}), solid at room temperature, derived from the higher-boiling petroleum fractions. There are three basic categories of petroleum-derived wax: **paraffin** (crystalline), **microcrystalline** and **petrolatum**. Paraffin waxes are produced from the lighter lube oil *distillates*, generally by chilling the oil and filtering the crystallized wax; they have a distinctive crystalline structure, are pale yellow to white (or colorless), and have a melting point range between 48°C (118°F) and 71°C (160°F). Fully refined paraffin waxes are dry, hard, and capable of imparting good gloss. Microcrystalline waxes are produced from heavier lube distillates and residua (see *bottoms*) usually by a combination of solvent dilution and chilling. They differ from paraffin waxes in having poorly defined crystalline structure, darker color, higher *viscosity*, and higher melting points--ranging from 63°C (145°F) to 93°C (200°F). The microcrystalline grades also vary much more widely than paraffins in their physical characteristics: some are ductile and others are brittle or crumble easily. Both paraffin and microcrystalline waxes have wide uses in food packaging, paper coating, textile moisture proofing, candlemaking, and cosmetics. Petrolatum is derived from heavy residual lube stock by propane dilution and filtering or centrifuging. It is microcrystalline in character and semi-solid at room temperature. The best known type of petrolatum is the "petroleum jelly" used in ointments. There are also heavier grades for industrial applications, such as corrosion preventives, carbon paper, and butcher's wrap. Traditionally, the terms *slack wax*, and *refined wax* were used to indicate limitations on oil content. Today, these classifications are less exact in their meanings, especially in the distinction between slack wax and scale wax. For further information relating to wax, see *blocking point, gloss, laminating strength, melting point of wax, oil content of petroleum wax, paraffin wax, petrolatum, refined wax, scale wax, scuff resistance, sealing strength, slack wax, strike-through.*

wax appearance point (WAP)--temperature at which wax begins to precipitate out of a *distillate* fuel, when the fuel is cooled under conditions prescribed by test method ASTM D 3117. WAP is an indicator of the ability of a distillate fuel, such as *diesel fuel*, to flow at cold operating temperatures. It is very similar to *cloud point.*

way--longitudinal surface that guides the reciprocal movement of a machine part. See *stick-slip motion.*

weed killer--*pesticide* product, often derived from petroleum, used for destroying weeds by direct application to the plants. "Selective" weed killers are those that, when applied in accordance with instructions, are not destructive to specified crops. Non-selective weed killers kill all vegetation.

weld point--the lowest applied load in kilograms at which the rotating ball in the Four Ball EP test either seizes and welds to the three stationary balls, or at which extreme *scoring* of the three balls results. See *four-ball method.*

wellbore--see *borehole.*

wellhead--the point at which oil and gas emerge from below ground to the surface; also, the equipment used to maintain surface control of a well.

wet gas--natural gas containing a high proportion of hydrocarbons that are readily recoverable from the gas as liquids. See *natural gas liquids.*

white oil--highly refined *straight mineral oil*, essentially colorless, odorless, and tasteless. White oils have a high degree of chemical stability. The highest purity white oils are free of unsaturated components (see *unsaturated hydrocarbon*) and meet the standards of the United States Pharmacopeia (*USP*) for food, medicinal, and cosmetic applications. White oils not intended for medicinal use are known as **technical white oils** and have many industrial applications--including textile, chemical, and plastics manufacture--where their good color, non-staining properties and chemical inertness are highly desirable.

wide cut--see *distillation test*.

wildcat--exploratory oil or gas well drilled in an area not previously known to be productive.

wood alcohol--see *methanol*.

worked penetration--see *penetration (grease)*.

workover--well that has been put back into production following remedial actions (such as deepening, clearing, *acidizing*) to restore or increase production.

worm gear--see *gear*.

xylene--aromatic hydrocarbon, C^8H^{10}, with three *isomers* plus ethylbenzene. It is used as a *solvent* in the manufacture of *synthetic rubber* products, printing inks for textiles, coatings for paper, and adhesives, and serves as a raw material in the chemical industry.

yield point--the minimum force required to produce flow of a plastic material.

ZDDP (zinc dialkyl dithiophosphate or zinc diaryl dithiophosphate)--widely used as an anti-wear agent in motor oils to protect heavily loaded parts, particularly the valve train mechanisms (such as the camshaft and cam followers) from excessive wear. It is also used as an anti-wear agent in *hydraulic fluids* and certain other products. ZDDP is also an effective *oxidation inhibitor*. Oils containing ZDDP should not be used in engines that employ silver alloy bearings. All car manufacturers now recommend the use of dialkyl ZDDP in motor oils for passenger car service.

ZN/P curve--general graphic representation of the equation: $\mu = (f)$ ZN/P, where μ (the *coefficient of friction* in a *journal* bearing) is a function (f) of the dimensionless parameter ZN/P, (viscosity x speed)/pressure. This is the fundamental lubrication equation, in which the coefficient of friction is

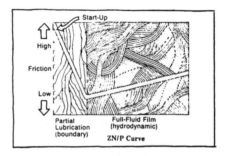

the friction per unit load. Z the viscosity of the lubricating oil, N the rpm of the journal, and P the pressure (load per unit area) on the bearing. The ZN/P curve illustrates the effects of the three variables (viscosity, speed, and load) on friction and, hence, on lubrication. See *boundary lubrication, full-fluid-film lubrication*.

INDEX

A

S